Environmental Footprints and Eco-design of Products and Processes

Series Editor

Subramanian Senthilkannan Muthu, Head of Sustainability - SgT Group and API, Hong Kong, Kowloon, Hong Kong

Indexed by Scopus

This series aims to broadly cover all the aspects related to environmental assessment of products, development of environmental and ecological indicators and eco-design of various products and processes. Below are the areas fall under the aims and scope of this series, but not limited to: Environmental Life Cycle Assessment; Social Life Cycle Assessment; Organizational and Product Carbon Footprints; Ecological, Energy and Water Footprints; Life cycle costing; Environmental and sustainable indicators; Environmental impact assessment methods and tools; Eco-design (sustainable design) aspects and tools; Biodegradation studies; Recycling; Solid waste management; Environmental and social audits; Green Purchasing and tools; Product environmental footprints; Environmental management standards and regulations; Eco-labels; Green Claims and green washing; Assessment of sustainability aspects.

More information about this series at http://www.springer.com/series/13340

Subramanian Senthilkannan Muthu
Editor

COVID-19

Sustainable Waste Management and Air
Emission

 Springer

Editor
Subramanian Senthilkannan Muthu
Head of Sustainability
SgT Group and API
Kowloon, Hong Kong

ISSN 2345-7651 ISSN 2345-766X (electronic)
Environmental Footprints and Eco-design of Products and Processes
ISBN 978-981-16-3858-9 ISBN 978-981-16-3856-5 (eBook)
https://doi.org/10.1007/978-981-16-3856-5

This Springer imprint is published by the registered company Springer Nature Singapore Pte Ltd.
The registered company address is: 152 Beach Road, #21-01/04 Gateway East, Singapore 189721, Singapore

Contents

About the Editor

Dr. Subramanian Senthilkannan Muthu currently works for SgT Group as Head of Sustainability and is based out of Hong Kong. He earned his Ph.D. from The Hong Kong Polytechnic University and is a renowned expert in the areas of environmental sustainability in textiles & clothing supply chain, product life cycle assessment (LCA), ecological footprint and product carbon footprint assessment (PCF) in various industrial sectors. He has five years of industrial experience in textile manufacturing, research and development and textile testing and over a decades of experience in life cycle assessment (LCA), carbon and ecological footprint assessment of various consumer products. He has published more than 100 research publications, written numerous book chapters and authored/edited over 100 books in the areas of carbon footprint, recycling, environmental assessment and environmental sustainability.

Triple Challenges of Sustainable Development: COVID-19, CO$_2$ Emissions and Public Debts—Findings from a Stylized CGE Model

Holger Schlör⬤ and Stefanie Schubert⬤

Abstract The current European situation can be characterized by a triple challenge: an acceleration in climate change and its consequences (a persistent drought in Europe since 2018), the deadly toll of COVID-19 and the rising public debts to cope with the COVID-19 consequences. These challenges endanger sustainable development in the European Union.

The Corona pandemic and the European governmental measures to contain the virus have serious economic and societal implications. The economic growth in the European Union has collapsed, consumption expenditures have significantly declined and the public fiscal budget deficit has risen sharply.

On a global scale, the Coronavirus epidemic has also increased the public debts worldwide. Financial debts are not the only debts the global community has to face. CO$_2$ debts denote debts of developed countries to developing countries for the damages caused by their disproportionate contributions to climate change. However, to meet the goals of the Paris Agreement, CO$_2$ emissions have to be reduced significantly to achieve the reduction goal for 2050 (UNFCCC in 2015, [93]; United Nations, Paris agreement. New York, 2015a, [94]). To accomplish this goal based on the recommendations of the Stern Report (Stern in The economics of climate change, The Stern review. Cambridge University Press, Cambridge, 2006 [85]), Metcalf (Metcalf in Paying for pollution: why a carbon tax is good for America, Oxford University Press (in production), Oxford, 2018 [57]) and the Journal Nature Climate Change (Nature Climate Change Editorial in Nat Clim Change 8:647, 2018 [61]), a CO$_2$ tax is being discussed.

Chapter to the book "COVID-19 & Sustainability" to be published by Springer-Nature Publications, editor: Dr. Subramanian Senthilkannan Muthu.

H. Schlör (✉)
Institute of Energy and Climate Research, IEK-STE, Forschungszentrum Jülinch, 52425 Jülich, Germany
e-mail: h.schloer@fz-juelich.de

S. Schubert
SRH Heidelberg, Ludwig-Guttmann-Straße 6, 69123 Heidelberg, Germany
e-mail: Stefanie.Schubert@srh.de

We take up these developments and analyze the socio-economic-ecological effects of the triple challenges Germany and the EU have to face. We use a computable general equilibrium (CGE) model that consists of two stylized EU countries with three economic sectors: A health sector, Food-Energy-Water-sector and a sector covering the rest of the economy (RoE). The model economy contains three economic agent groups: the private households, the firms and the governments.

The CGE model allows the macroeconomic impacts of the Corona pandemic to be analyzed. The concept of Lastenausgleich (burden sharing) is considered within the model to distribute the burdens of the Corona measures to mitigate the COVID-19 effects and the CO_2 debts. Therefore, a CO_2 tax is also implemented in the model.

Keywords COVID-19 · CO_2 emissions · Public debts · Sustainable development · Sustainable development goals · European union · CGE model · Lastenausgleich

1 Introduction

The Sars-CoV-2 pandemic started twelve month ago and has caused 2.7 million deaths. The virus impacts the core of the global society: It forces us to reshape our daily life,

- how we work,
- how the children work,
- how we organize the social life and
- how we die [76].

Examples for this new societal development are [76]:

- Portions of last year's French wine production have been turned into the industrial alcohol that restaurants offer their guests to disinfect their hands [76].
- The airline Quantas plans to carry only passengers vaccinated against COVID-19 on intercontinental flights.[1]
- The Spanish government wants to create a Europe-wide file of vaccination refusers.[2]
- The head of the German vaccine company Biontech expects a return to normalcy by Christmas–Christmas 2021.[3]
- Yuval Noah Harari fears that the pandemic may later prove to be the beginning of total surveillance of humanity [76].

---

[1]https://www.bbc.com/news/world-australia-55048438.

[2]https://www.bbc.com/news/world-europe-55471282.

[3]https://www.aerzteblatt.de/nachrichten/118377/Biontech-Chef-erwartet-Rueckkehr-zur-Normalitaet-Ende-naechsten-Jahres.

- And Melanie Amann wrote in the German magazine 'Der Spiegel' about the political situation in Germany: So many unbelievable mistakes have been made, annoyingly small ones and dramatically large ones, that one stands amazed and sometimes frightened in front of one's own country, which we thought we knew quite differently [2].[4]
- And David Malpass, president of the World Bank, describes the current situation: "More than a year into the COVID-19 pandemic, the scale of the tragedy is unprecedented [54]."
- The German President summed up the current state of society as follows: We are weary from the burden of the pandemic and chafed in the argument about the right way forward. Since the beginning of the catastrophe, we have been looking at infection rates and death figures every day as if spellbound. [But behind all the numbers are human fates.] Their suffering and deaths have often remained invisible to the public. A society that suppresses this suffering will suffer as a whole (Bundespräsident Frank-Walter [14].[5]

People lost their faith in national and international institutions because the Corona pandemic has shown how unresilient and vulnerable societies are regardless of their level of development. This was a shock especially for Western industrial countries [76].

In this chapter, we provide a brief overview of the global course of the Corona pandemic, its historical background, its macroeconomic impact and the resulting increase in public debt in the first section. In addition, Sect. 3 introduces the third global challenge: climate change and CO_2 debt. In Sect. 4, the German concept of burden sharing (Lastenausgleich) is presented to deal with these three challenges. In Sect. 5, we offer an overview of our CGE model. Finally, in Sect. 6, the model results are presented. Based on the results of the CGE model, Sect. 7 discusses how the Corona pandemic affects the UN Sustainable Development Goals. Concluding remarks are made in Sect. 8.

2 Covid-19

Hence, in the following, a more detailed picture of the challenges set by the Corona pandemic is presented, which have caused all the socio-economic distortions in the last year around the globe. The Corona pandemic is not the first health crisis that hit the global community. The current pandemic lines up in a long list of severe pandemics in the last 2000 years.

---

[4]Translated by the authors.
[5]Translated by the authors.

2.1   Historical Background

In the past, several examples of deaths from various pandemics were reported within just a few years. In the fourteenth century, 200 million people died globally due to the plague and 37% of the European population died [10, 15, 29]. The smallpox pandemic in the sixteenth century takes the lives of 56 million people [65], as Fig. 1 shows.

The Spanish flu pandemic (1918–19) was the most dangerous pandemic in the twentieth century according to the Centers for Disease Control and Prevention [16]. The CDC estimated that 500 million people were infected equaling one-third of world's population, killing more than 45 million people. The Spanish flu pandemic was characterized by the disproportionate deaths of young and healthy people in particular [16]. In the sixth century, the Justinianic Plague [100], the first part of the three plague pandemics, had devastating effects on the Mediterranean region. This plague caused 40 million casualties in a world population of 213 million [65, 82].

In 2019, the Joint United Nations Programme on HIV/AIDS (UNAIDS) estimated the number of people living with HIV/AIDS worldwide to be around 38 million. 690 000 people died from AIDS-related illnesses in 2019. 75.7 million people became ill since the start of the epidemic and 32.7 million people have died from AIDS-related illnesses since 1981 [92].

The three pandemics (Hongkong flu, Asian flu and COVID-19) of the last 70 years caused 4.3 million casualties, whereas as of March 2021, Corona has already caused more deaths than the Asian flu and the Hongkong flu altogether, see Fig. 1.

The Corona pandemic started at the end 2019.

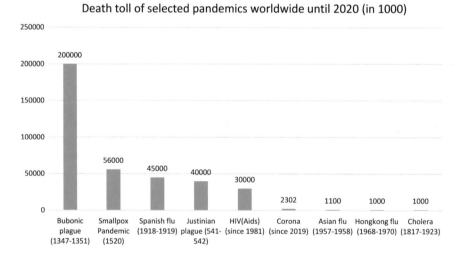

Fig. 1 *Source* Statista [80, 81]

2.2 Coronavirus

The government of Wuhan in China confirmed on December 31 the first cases of the new Coronavirus. On January 11, the Chinese government confirmed the first Corona death [62]. Two weeks later Wuhan was cut off by the Chinese government from the rest of China and one week later on January 30 the World Health Organization (WHO) declared a global health emergency [104]. On February 23, the first Europeans died of Corona in Italy and on March 19 Italy recorded more victims than China [19].

A pandemic is characterized by the German Robert-Koch-Institut (RKI) as a new, but temporary, strong worldwide spread of an infectious disease such as the Coronavirus with high numbers of cases and severe courses of disease. The WHO can declare a pandemic in accordance with the International Health Regulations [69].

On March 11, 2020, WHO Director-General Dr. Tedros Adhanom Ghebreyesus officially declared the COVID-19 outbreak a pandemic [102]. This decision was made due to the rapid increase in the number of cases outside of China [87].

On April 26, the global death toll surpassed 200,000 and in May 17 Japan and Germany entered an economic recession [17]. In July 21 the European leaders agreed on an $857 billion economic stimulus package because their economies were hit very hard by the pandemic [25]. On September 28, the global death toll of the virus reached 1 million victims [87].

But in the fall of 2020, Coronavirus mutations were detected including three mutations of concern [70]. Variant B.1.1.7 has been spreading in the UK since December 2020 [77]. The German Robert-Koch Institut (RKI) suspects that the UK variant could lead to more severe courses of disease [70] but not to increased mortality as recent studies show (Frampton et al., Graham et al.) The new variant caused about 28 percent of all cases in London in early November 2020 and 62% a month later [70]. Variant B.1.351 is also a mutation variant of the SARS-CoV-2 Coronavirus [22, 109]. This mutant was discovered in South Africa in December 2020 and reported by the country's Department of Health on December 18, 2020 [9, 108].

Variant B.1.1.28 [68] first appeared in the Brazilian state of Amazonas and is similar to the South African variant [91]. The RKI also suspects an increased risk of infection and an impairment of the effectiveness of vaccines [68].

2.3 Corona Cases

The cumulative number of confirmed SARS-CoV-2 infections worldwide is more than 132 million (as of April 9, 2021). The number of deaths associated with the Coronavirus increased to more than 2.8 million by that date [105]. The cumulative cases are distributed over the different regions as Fig. 2 shows. North America confirmed nearly 24% of the Corona cases worldwide and has to bear 20% of the global victims. In Europe, Africa and Western Pacific the share on the total cases and deaths are very similar, whereas Central, South America and the Caribbean are the region where the proportion of global deaths from the pandemic (28.2%) is much higher than the proportion of total Corona cases (19.5%).

North America and Europe account for more than 50% of the world's corona cases and deaths. Statista [81]. Figure 3 shows a more differentiated picture of the Corona pandemic.

Figure 3 shows that France has to face the highest cumulative Corona cases in Europe followed by Spain, Italy and Germany. Striking is that in the case of Italy the share (18.2%) on the EU27 confirmed Corona caused deaths is the highest in the European Union. A similar picture is revealed for Germany, where also the share on the EU27 death toll is higher than the share on the confirmed COVID-19 cases [23].

This relation is also visible for Poland, Romania, Beligum, Hungary, Bulgaria and Greece, i.e. in eight countries of the EU27 the share on the total deaths toll of COVID-19 is higher than the share on the Corona cases. Figure 4 shows exemplary for Germany how the Corona victims are distributed between the different age groups.

In Germany, 80% of deaths among women are attributable to the age group 80 and older, while in the corresponding age group among men this age group accounts for 61% of all deaths. In the other age groups, male victims are over-represented. In the under-60 age group, the death rate for German men was 4.5% and for women 2.1%. The dangerousness of COVID-19 increases significantly with the age. The WHO sees in the vaccination the only effective and safe way to stop the Corona pandemic and save lives [101] especially in older age groups [73].

2.4   COVID-19 Vaccines

The rollout of the COVID-19 vaccines started in the European Union in December 27, 2020 [18].

Figure 5 reveals the vaccination situation in February 2021 and shows that only 1.25% of the World population was vaccinated for the first time and only 0.41% of the World population has received the second vaccination for full protection, i.e. 99.59% of the population had no vaccination protection at all [1]. Figure 5 also shows how unevenly vaccination is distributed across the global population. Israel has achieved full protection of one-third of the population followed by Seychelles and the Cayman Islands. The USA has already achieved full protection for 4.63% of its population, whereas the European Union has only reached full coverage for 1.77% of its population. Striking is that the UK has vaccinated for the first time 23.5% of its population but only 0.82% received a full coverage of two vaccinations.

The Secretary-General of the United Nations Antonio Guterres appreciates the rollout of the COVID-19 vaccines. But he stressed that the global community is at a critical moment because it faces a new moral challenge: the community must ensure that everyone can get a vaccination [39]. However, the current distribution of

Covid-19 confirmed cases and deaths - in % on total cases April, 9, 2021

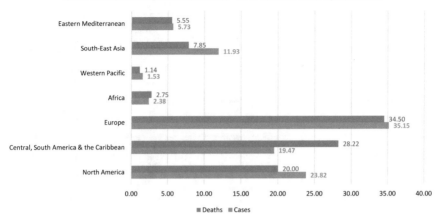

Fig. 2 *Source* World Health Organization (WHO) [105] and own calculations, 2021

Covid-19 confirmed cases and deaths in EU 27 until February, 7, 2021 - in % on total EU

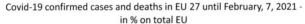

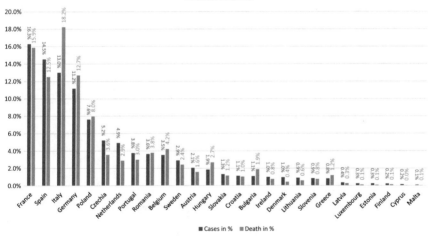

Fig. 3 A more differentiated picture of the Corona pandemic. *Source* Own calculation, 2021, based on European Centre for Disease Prevention and Control [23]

vaccines is very unfair, as 10 countries keep 75% of the vaccines produced. The global community will only be protected if everyone is vaccinated [39].

The WHO, European Commission, France and The Bill & Melinda Gates Foundation set up in April 2020 the Access to COVID-19 Tools (ACT) Accelerator [103] to end the COVID-19 pandemic [24, 27]. COVAX (COVID-19 Vaccines Global Access) is the central pillar of the Accelerator to deal with the pandemic [50, 88]. It aims at providing COVID-19 vaccines around the world [31]. The German

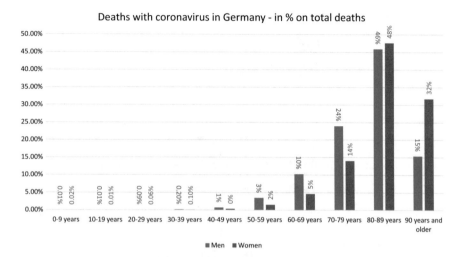

Fig. 4 *Source* Robert-Koch Institut (RKI) [71] and own calculations

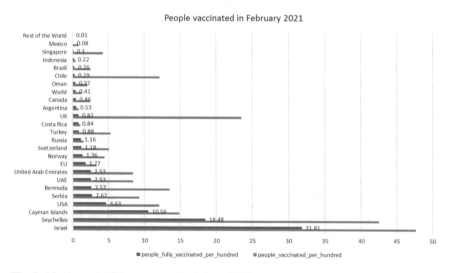

Fig. 5 Ritchie et al. [67], and own calculations, 2021

government supports the COVAX alliance and the German Foreign Office stressed that vaccines must therefore be accessible and affordable worldwide (Auswärtiges Amt der Bundesrepublik Deutschland [6]. This is the goal of the COVAX platform, led by the Gavi vaccine alliance, the World Health Organization (WHO) and the Coalition for Epidemic Preparedness Innovations (CEPI[6]). Until February 2021, 97

---

[6]https://cepi.net/.

industrialized and emerging countries, including Germany, have joined COVAX initiative but still, 130 countries have no access to the new vaccines [31].

The WHO Director-General Tedros Ghebreyesus stressed also the need for a fair distribution of the vaccines because the uneven distribution of vaccines would only prolong the pandemic unnecessarily and increase human suffering and economic difficulties. An equal distribution of vaccines is not only an ethical question in the sense of the categorical imperative [48], but also a question of economic reason [34].

He sees the danger that the vaccines will deepen the inequality between those who have access to vaccines and the poor rest of the world. This inequality gap is expressed in the fact that "just 25 doses have been given in one lowest-income country. Not 25 million; not 25 thousand; just 25 [34]."

Ghebreyesus stressed also that the equal distribution of vaccines is not only a moral issue but has also economic benefits for high-income countries [37]. The benefit for ten high-income countries would outweigh the cost of equitable distribution of vaccines [34].

Hence, we have so far characterized the public health situation caused by the Coronavirus. In the following, we will take up the considerations of Ghebreyesus and focus on the macroeconomic effects of the Corona pandemic.

2.5 Corona Macroeconomic Impact

The spread of the SARS-CoV-2 and the governmental measures to contain the pandemic have serious economic and societal implications, as Fig. 6 shows [84].

Between the first quarter of 2018 and the last quarter of 2019, the GDP in the EU27 grows on average about 0.525%. The decline starts in the first quarter of 2020, when the economic performance decreases about −3.3% and in the second quarter the decrease accelerated to -11.4% for the European Union. In the first two quarters, the European Union economy contracted by −14.7%. In the third quarter, the EU economy was able to recover by 11.% but it remains a rate of decline of - 3.2% over the three quarters of 2020. In the first two quarters of 2020, the biggest slump of the GDP has taken place in Spain (−23.2%), followed by France (−19.6%), Malta (−19.8%) and Italy (−18.5%), whereas a moderate decrease takes place in Finnland (−5.4), Lithuania (−5.9%), Estonia (−6.3%) and Ireland (−6.7%).

The decline of the GDP is also correlated to the change in consumption expenditures, as Fig. 7 shows. Consumption expenditures decline in the EU27 by about −16.1% in the first two quarters in 2020 and increased in the third quarter about 13.2%, so that in the first three quarters the consumptions declines about 2.9%.

The highest decline occurs also in Spain, where the consumption decreases in the first two quarters about 26.9%, followed by Ireland (−22.7%), Malta (−22.5%) and Luxemburg (−20.7%), whereas Bulgaria measured only a small decline of −2.7%, followed by Slovakia (−6.1%) and Estonia (−7.3%).

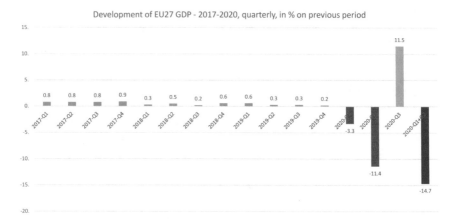

Fig. 6 *Source* Eurostat [26], and own calculations

The decline in GDP and consumption in the EU has also an impact on the overall economic and business climate of the European Union. In the European Union, the business climate declines from 103.6 points in February 2020 to 67.1 points in April 2020. The indicator lost 36.5 points in two months. This was the sharpest decline since the survey began in 1985 [84].

The decline in GDP in European Union forced the European Commission to rise the budget deficit to €1 trillion to counter the Coronavirus crisis [4]. According to calculations by the Financial Times, this would lead to a huge increase in the budget deficit [4]. A closer look on the economic conditions of the countries of the Eurozone reveal the financial burden caused by COVID-19.

Germany and France are covering nearly 45% of the Eurozone budget deficit, followed by Italy and Spain (18.2%, 12.8%). These four countries are covering 75.5% of the deficit of the Eurozone fiscal budget deficit. The deficit of the Eurozone amounts to €976 billion [4]. But Fig. 8 also shows that the budget deficit in euro per inhabitant shows a different picture. The average financial burden for the inhabitants varies significantly between the countries of the Eurozone.[7]

The average inhabitant of the Eurozone has to bear a financial burden of €2847.2 due to the financial measures in the Eurozone countries to defy the Coronavirus. The highest financial burden of €7028 affects the inhabitants of Luxembourg followed by Austria, Ireland, Belgium, France and the Netherlands. The lowest burden has to bear Cyprus, Latvia and Estonia. Luxembourg's financial burden is more than 8 times that of Cyprus.

The Biden administration will spend $1.9 trillion with the goal to reach full employment and wage growth [63]. The American Rescue Plan [89] includes a $1,400-per-person check for most American households ($400 billion), an increase

---

[7]The Eurozone is the monetary union of 19 member states of the European Union (EU) that have adopted the euro (€) as their primary currency.

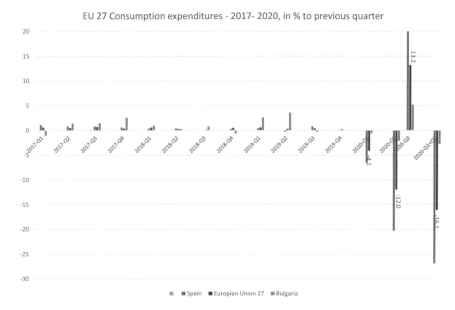

Fig. 7 *Source* Eurostat [26], and own calculations

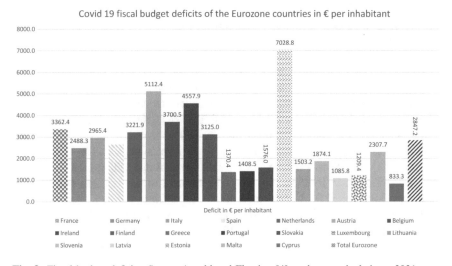

Fig. 8 Fiscal budget deficits. *Source* Arnold and Fleming [4], and own calculations, 2021

of unemployment benefits of $240 billion; an increase in tax credits for parents as well as low-income employees of $130 billion; further, health insurance benefits of $65 billion; as well as housing assistance of $40 billion [64, 89].

Table 1 Comparing Debt-to GDP by economic sector, Q3, 2019 to Q3, 2020

	Households	Non-financials* Q3, 2019 = 100	Governments	GDP Growth reduction in %
Developed markets average	108	112	119	0.39
USA	109	117	125	0.66
Euro Area	105	106	113	0.32
UK	105	107	118	0.43
Emerging markets average	110	112	113	0.19
China	111	111	119	0.27
Russia	121	117	120	0.22
Global total	108	112	118	0.46

Source Marcus Lu, 2021 and own calculations, 2021, *corporations not in financial industry

All those governmental activities to mitigate the global economic impact of the pandemic result in an increase in global debts "by $20 trillion since the third quarter of 2019 [53]." It is assumed that the global debts rise to $277 trillion by the end of 2020 corresponding to 365% of world GDP, as the Institute of International Finance calculated.[8] Table 1 shows how debt has changed in the main economic sectors: Households, non-financial sector and governments [53].

The debt-to-GDP ratio increased for the households of the Euro Area about 5 percentage points, for the non-financial corporations about 6 percentage points and rise for the governments of the Euro Area about 13 percentage points in just one year as Table 1 shows. The highest increase occurs in the private households in Russia (21 percentage points) and China (11 percentage points). The debt burden of the US households increased about 9 percentage points and about 12 percentage points for the non-financial corporations. The governmental debt burden rises about 25 percentage points so that the debt-to-GDP burden of the US government is now 127%, for the EU area 115% and for the UK 130%. This debt increase has also a negative impact on the future economic growth development, as the World Bank detected [36]. The World Bank was able to show that this will reduce the growth rate in the Eurozone by 0.32 percentage points and in the US and UK by 0.66 percentage points and 0.43 percentage points respectively [36].

The analysis shows how the corona pandemic increased the global debts and how this increase will reduce future GDP growth (Table 1).

3　Climate Change

The Coronavirus epidemic increased the public debts globally, but the financial debts are not the only debts the global community has to face [93, 94].

---

[8]https://www.iif.com/COVID-19.

3.1 CO$_2$ Debts

The CO_2 debts refers to the debts that developed countries owe developing countries for the damages caused by their disproportionate contributions to climate change [44, 55]. The cumulative greenhouse gas emissions, largely caused by developed countries, also pose a significant threat to developing countries, which, however, are much less able to cope with the negative impacts of climate change, as the following table shows [44, 66].

According to Table 2, the USA has emitted 410.24 billion tons of CO_2 between 1751 and 2011 and causes 27.3% of all CO_2 emitted since 1751, followed by China, Russia, Germany, UK, Japan and France (14.7, 7.6, 6.1, 5.2, 4.3, 2.6%). Table 2 also shows the contribution of these emissions to the global temperature increase. The USA are responsible for 35% of the global temperature increase, followed by Russia and China. Germany, the UK and France, as the main contributors to the CO_2 emissions of the European Union, are responsible for 19.6% of the global temperature increase. Table 2 also shows that only 20 countries have caused 82.5% of the cumulative CO_2 emissions since 1751.

Figure 9 shows now the accumulated CO_2 debts the current generation has to bear. Every citizen of the USA stands in 2011 for 1270 tons of CO_2 that the USA have emitted since 1751, followed by the United Kingdom (1179.4 t CO_2), Germany (1108.2 t CO_2), and Canada (871.3 t CO_2).

3.2 Rising Global Temperature

Figure 10 reveals the increase of the global surface temperature relative to the 1951–1980 average temperatures caused by the cumulative CO_2 emissions. The global surface temperature increased about 1.02 °C in 2020, as measured by NASA [59].[9]

Against this background, the idea of CO2 debts refers to the fact that the countries contribute differently to the current stock of CO_2 emissions since 1751 [12, 55]. The historical global emissions of greenhouse gases are largely caused by the developed countries, but pose a serious threat to the rest of the world [12]. The historical CO_2 debts also raise questions of equity [58] and of responsibility: "who owes whom for what [12]."

The main components of CO_2 debts are adaptation and emission debts. Adaptation debts refer to the debts that industrialized countries owe to developing countries to help them financially to better adapt to climate change. Emission debts refer to the debts that industrialized countries have to bear for their disproportionate greenhouse gas emissions [12].

---

[9]https://climate.nasa.gov/.

Table 2 Cumulative CO_2 emissions from 1751 to 2011

Emissions since 1751			Contribution to global temperature change		
In billion tonnes	In % total	Population in billions	Emission per capita in tonnes	In °C	In %
United States	410.24	27.3	0.323	1270.1	0.143	35.0
China	219.99	14.7	1.400	157.1	0.042	10.3
Russia	113.88	7.6	0.144	790.8	0.059	14.4
Brazil	15.13	1.0	0.210	72.0	0.004	1.0
India	51.94	3.5	1.352	38.4	0.013	3.2
Germany	91.98	6.1	0.083	1108.2	0.035	8.6
UK	77.84	5.2	0.066	1179.4	0.031	7.6
France	38.26	2.6	0.067	571.0	0.014	3.4
Indonesia	13.5	0.9	0.270	50.0	0.003	0.7
Canada	33.11	2.2	0.038	871.3	0.011	2.7
Japan	64.58	4.3	0.127	509.1	0.021	5.1
Mexiko	19.76	1.3	0.128	154.9	0.006	1.5
Thailand	7.15	0.5	0.070	102.7	0.002	0.5
Columbia	3.34	0.2	0.050	66.4	0.001	0.2
Argentina	8.29	0.6	0.045	184.6	0.002	0.5
Poland	27.56	1.8	0.038	718.6	0.01	2.4
Nigeria	3.82	0.3	0.214	17.9	0.001	0.2
Venezuela	7.74	0.5	0.029	271.6	0.002	0.5
Australia	18.18	1.2	0.025	715.7	0.005	1.2
Netherlands	11.63	0.8	0.017	668.4	0.004	1.0
Sum	1237.92	82.5	4.7	263.6	0.409	100.0
World 1751	1500	100	7	214.3		

Source Höhne et al. [44], Ritchie [66], and own calculations, 2021

3.3 First Summary

The Corona pandemic is not the first health crisis that hit the global community. The current pandemic lines up in a long list on severe pandemics in the last 2000 years. The Corona pandemic has broken out in Wuhan, China and is characterized by the WHO as a new pandemic. SARS-CoV-2 infections worldwide have increased to more than 105.4 million by Feb. 9, 2021. The number of deaths associated with the Coronavirus increased to more than 2.3 million by that date (WHO). Europe and North America account for 50% of the World's Corona cases and deaths. The governmental measures to contain the Corona pandemic have caused a decline of the economic growth, a break-in of consumption and a rise of the public fiscal debts. The rollout of the COVID-19 vaccines started in December 2020 but the WHO sees the danger that the current vaccination will deepen the inequality between those who have access to vaccines and the rest of the world.

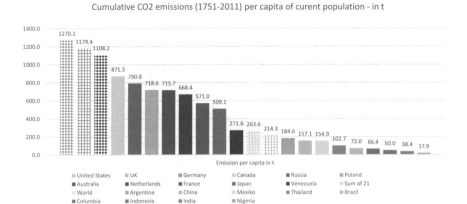

Fig. 9 *Source* Own calculations based on [44, 66]

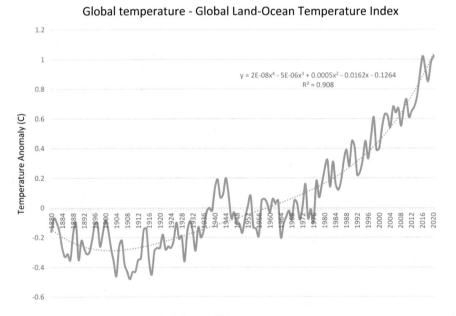

Fig. 10 NASA [60] and own calculations, 2021

In summary, then, the current trend in global debts highlights the urgency of new institutional measures to better share the fiscal burden, to achieve greater reductions in greenhouse gas emissions and to reduce the economic pressure of rising fiscal debts on future generations.

4   Burden Sharing (Lastenausgleich)

In the following, the idea of Lastenausgleich (the burden sharing of economic pressure) is presented to deal with the economic challenges of today's severe crises and its costs: Corona pandemic and climate change. The idea of Lastenausgleich was developed after the Second World War in Germany (1948/49) [45] and further developed after the fall of the Berlin Wall (1990/91) in the course of the German reunification [41] and is now being discussed to better share the costs of the Corona pandemic (2020/2021) [52]. It could also be an instrument to bear the costs of climate change [12, 61, 75].

4.1   Lastenausgleich—Historic Models

Those responsible in the American and British occupation zones after the Second World War had recognized early that in view of the economic and social situation in West Germany at the time, the integration of millions of destitute people would not be possible without administrative measures [13]. Against this background, the government of the Federal Republic of Germany supported by the Western Allies adopted the first Lastenausgleichsgesetz (equalization of burdens law) in 1952[10] [8, 45, 98], which was intended to socially equalize the unequally distributed burdens of the consequences of the Second World War [52]. The aim of the law was also to develop a compensation model to offset the undesirable economic effects of the currency reform of 1948. The currency reform devalued savings, while investors in real estate or gold hardly suffered any losses and borrowers even benefited from the devaluation of their debts [52].

The Lastenausgleich law of 1952 had three parts: a levy into a property fund, a mortgage profit levy and a credit profit levy. The levy into the property fund generated revenues of €21.5 billion, the mortgage profit levy created revenues of about €4.45 billion and the credit profit levy yielded €0.92 billion. Additionally, the federal and state governments supported the Lastenausgleich fund with €31.2 billion between 1949 and 2001. So that the fund administered €58 billion to support the German population [41].

After the fall of the Berlin wall and the reunification of Germany a second Lastenausgleich project was established, which was not named like this [41]. The Lastenausgleich leads to various redistributions by the German government and the German social security system. These institutions organized a redistribution to the East German institutions and households [41].

To finance the redistributions a solidarity surcharge was adopted by the German parliament in 1995.[11] From 1995 to 1997, the surcharge was 7.5%; since 1998, it

---

[10]https://www.gesetze-im-internet.de/lag/BJNR004460952.html.

[11]https://www.gesetze-im-internet.de/solzg_1995/index.html.

has been 5.5% of the paid tax. It is levied on all taxpayers through a surcharge on income tax, wage tax, capital gains tax and corporate income tax [41, 75].

To finance German unification, the solidarity surcharge generates tax revenues of €345.9 billion between 1991 and 2019 (see Fig. 11) [42, 83] Germany uses again the idea of the Lastenausgleich to finance the extraordinary challenges of the unification.

Hence, we can summarize so far the idea and the advantages of the Lastenausgleich [41]:

1. First, a balance can be set between the population groups which are more and or less burdened by unpredictable great challenges;
2. Second, the avoidance of impoverishment of large population groups; and
3. third, the creation of a social structure that stabilizes the democratic political system and promotes economic development.

The question now is, whether the idea of Lastenausgleich idea is applicable to the challenges of CO_2 debts.

4.2 Lastenausgleich Climate Change—Carbon Pricing

To address rising CO_2 emissions, Nicolas Stern describes climate change as a global market failure in the Stern Report and supports as a central policy element the pricing of carbon [85]. The CO_2 tax can encounter the global threats of climate change [57]. Metcalf [57] recommended in his book: *Paying for Pollution: Why A Carbon Tax is Good for America* also a CO_2 tax. A CO_2 tax is also recommended

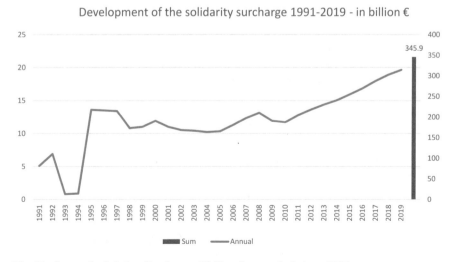

Development of the solidarity surcharge 1991-2019 - in billion €

Fig. 11 *Source* Statistisches Bundesamt, BMF and own calculations, 2021

by the Journal Nature Climate Change [61]. In Europe, several models of CO_2 tax have been implemented [5]. The Swedish households and services pay 108.81 euros per ton of CO_2 (April 1, 2020). In France, a carbon tax was implemented in 2014. The tax was €44.81/ton of CO_2 in 2020 and will increase to €86.20/ton in 2022. In 1999, Germany implemented an ecological tax reform, which was adjusted in 2000 and in 2003. In December 2019, the German government agreed on a carbon tax of 25 euros per ton of CO_2, which came into force in January 2021. By 2025, the tax rate will be increased up to 55 Euros per ton of CO_2 [110].

4.3   Second Summary

The idea of Lastenausgleich developed in Germany (1948/49) [45] and was further developed in the course of the German reunification [41], discussed as a model for carrying the Corona pandemic (2020/2021) costs [52]. It could also be an instrument to bear the costs of climate change [12, 61, 75] as developed by David Stern and Metcalf [56, 57, 85].

We take up the idea of the Lastenausgleich in our model to better redistribute the economic burdens and thus avoid transferring the financial burdens of the Corona pandemic and of climate change to future generations through additional debts. We, therefore, introduce a CO_2 tax in our model economy. The CO_2 tax revenues are redistributed to households to compensate them for the loss of utility caused by the Corona measures.

5   The Model

For our analysis, we developed a dynamic General equilibrium model [40, 72]. Computable General Equilibrium (GCE) models [20] are based on neo-classical economic theory [78, 90] and are used for macro- and microeconomic analysis [11]. They enable to define the endogenous variables of the model economy and can compute the structural parameters of the model economy [38, 96]. Thereby, the CGE models enable the analysis of economic adjustment effects caused by external shocks. The model enables to estimate how the economy reacts on policy changes such as the governmental regulations for the containment of the Corona pandemic [49, 99] or through new CO_2 taxes [7]. The advantage of the CGE model is that not only direct economic effects are considered, but also indirect feedback effects can be taken into account.

Our model economy consists of two model EU countries and covers four periods. This allows us to concentrate on the direct economic effects arising from the Corona pandemic instead of including other side effects, e.g. the EU debt crisis. The model framework is similar to the Ecomod model [21].

5.1 The Model Framework

The economies of the two EU countries consists of three sectors. Each sector is represented by a firm, which operates under perfect competition and constant returns to scale with Cobb–Douglas technology. For the consumers, it is assumed a rational forward-looking expectations [30, 107], with no money illusion [110] and no uncertainty about the future. The producer has to determine the optimal labour input and investment in each period of time. It is also assumed that the producers have a rational forward-looking expectation [30, 107]. There is no uncertainty, no money illusion [110]. Thus, the producer decides on labour and capital use at each time period but also on investment in order to maximize the value of the firm. The governments of the two EU countries collect the Lastenausgleich taxes and compensate households by transfers.

The two EU countries trade with each other. For the substitution of goods produced domestically by goods produced abroad, we apply the Armington assumption. [3]. It implies that goods sold domestically but produced in different countries exhibit elasticity of substitution and thus goods are differentiated according to their origin. The Armington assumption can be seen as a special case of horizontal product differentiation [74, 106].

5.2 Social Accounting Matrix

The Social Accounting Matrix (SAM) covers the data that is used for calibrating our theoretical model. The SAM is a structured framework that allows capturing the economic transactions of an economy [97]. The UN stresses the accounting consistency and comprehensiveness of the SAM for economic analysis. In the assessment of the UN, the SAM is the preferred tool used to calibrate computable general equilibrium (CGE) models [96].

The model economy consists of two EU countries A and B. Country A represents a smaller EU country and country B is representing a larger EU country. Every country consists of three economic sectors: a health sector, a Food-Energy-Water (FEW) sector and a sector that represents the rest of the economy (RoE).

Table 3 represents the Social Accounting Matrix of the smaller EU country A. The SAM contains the initial capital and labour demand of the country, gross output, investments and the imports and exports of country A.

Table 4 represents the Social Accounting Matrix of the greater EU country B. The SAM consists also of three economic sectors, the gross output and the investments at the starting period and captures the trade relations of country B.

We assume for the model, that country A has a CO_2 intensity of 0.23. The CO_2 intensity measures the ratio of carbon dioxide emissions to the gross domestic product. The value of 0.23 is oriented on the CO_2 intensity of the average OCED country in 2019, whereas it is assumed for country B a CO_2 intensity of 0.172. This value is based on the CO_2 intensity of the average EU28 country also in 2019 [46].

Table 3 Social accounting matrix EU country A-in currency units

	Health	FEW	RoE	Consumption	Investment	Exports	Total
Health	0	0	0	245	17.5	52.5	315
FEW	0	0	0	420	140	70	630
RoE	0	0	0	647.5	350	175	1172.5
Capital (K) payments	175	210	350				
Labour (L) payments	70	315	700				
Gross output (XD)	245	525	1050				
Imports	70	105	122.5				
Total	315	630	1172.5				

Source ECOMOD [21] & authors, 2021

Table 4 Social accounting matrix EU country A-in currency units

	Health	FEW	RoE	Consumption	Investment	Exports	Total
Health	0	0	0	437.5	35	70	542.5
FEW	0	0	0	735	280	105	1120
RoE	0	0	0	1452.5	700	122.5	2275
Capital (K) payments	350	420	700				
Labour (L) payments	140	630	1400				
Gross output (XD)	490	1050	2100				
Imports	52.5	70	175				
Total	542.5	1120	2275				

Source [21] and authors, 2021

We further assume a CO_2 tax. The CO_2 tax taxes the CO_2 emissions of the gross output and the CO_2 emissions of the consumption. The CO_2 tax is 0.025 monetary units per kg CO_2 for EU country A and 0.11 monetary units per kg CO_2 for country B. The taxes will be redistributed to the households of the two countries. It is assumed further a similar interest and time preference rate for both countries of 0.002%.

Additionally, a steady-state growth rate of -10% for country A and -2.5% for country B is assumed to capture the heterogenous development in the European Union presented in Fig. 6.

6 Results

The following chapter presents the results of the key indicators of the CGE model. Nine indicators have been selected to present the results of our model economy consisting of two EU countries.

6.1 Income and Savings [Y,S]

In the two-country-model economy, the Corona-induced economic slump causes for the large EU economy B a decline of the income of about 7% in the analyzed period, whereas the small EU Country A is affected by a 27% income reduction.

Figure 12 reveals the spread of the income distribution caused by the simulation of the economic effects of the Corona pandemic. The savings of the two EU model countries also decline in the analyzed 4-year period. The savings of country B decreases proportional to the income, whereas the savings of the small country A decrease a little bit less.

The part of the income which is not saved for future investments and savings is used for consumption.

6.2 Consumption [Con]

Figure 13 shows the development of the consumption in the two EU countries. The analysis reveals that the consumption of the health goods decrease in EU country A in the analyzed period of about 62.9 monetary units and the consumption of the FEW-nexus sector goods declines about nearly 110 monetary units. The consumption of the goods of the rest of the economy declines about ca. 170 units. The consumption of country A break-in about 341 nominal monetary units.

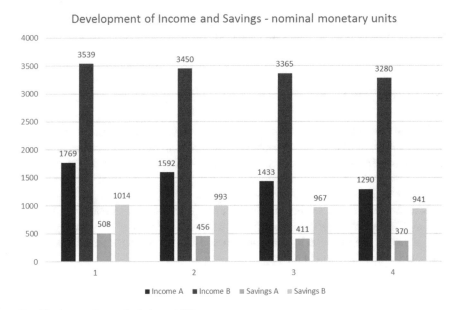

Fig. 12 *Source* Own calculations, 2021

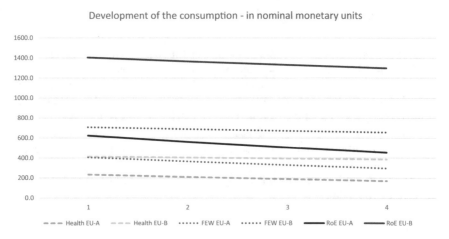

Fig. 13 *Source* Own calculations, 2021

Figure 13 reveals also that the economy of country B is struggling with a milder decline in consumption. The weakest consumption decrease takes place in the health sector (−29.2 monetary units), followed by the goods of the rest of the economy (−105.5) and the FEW sectors (−51.1). The consumption of country B decreases about 186 monetary units. The households of the large country B also renounce the consumption of the commodities of the FEW sector and of the rest of the economy (RoE) sector to avoid a steeper reduction of the consumption of the health commodities.

The decline in consumption and savings also has an impact on investment in the two countries, as the following figure shows.

6.3 Investments [Inv]

It can also be noted that the two model countries respond differently to the economic challenges of the Corona pandemic with their investment decisions (Fig. 14). Country A—the small EU country—reduces its health investments in the analyzed period by about 32.8 monetary units. Country B's investment in this sector decreases by only 15.6%.

Investment in the other sectors also declines in both countries over the observed period. The decline in EU country A is always steeper than in EU country B.

Inevitably, the changes in consumption and investment have an impact on the gross output of the two countries.

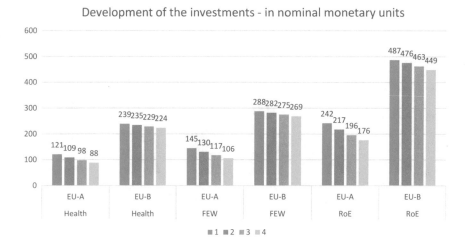

Fig. 14 *Source* Own calculations, 2021

6.4 Gross Output [GO]

The gross output (GO)—the measure for the total economic activities (production of new goods and services) of the two model EU economies—reveals also a different development in the two countries and their sectors (Fig. 15).

The gross output of the health sector of country A decreases slightly about 67 real good units to meet the challenges of the modelled Corona pandemic economic environment. The output of the FEW sectors of EU country A declines about 143 units, whereas the gross output of the RoE sector decrease about 27%.

Country B's health goods output declines by only about 37 units over the entire 4-year period, while EU country B's FEW sector output declines by more than 80 units and the rest of the economy's output declines by only about 7%. Thus, the total output of the two countries falls by 759 real goods units.

In the following, it will be analyzed how the so-far described economic development affects the labour market of the two countries.

6.5 Labour Demand [La]

The labour demand decreases in all three economic sectors of the two economies about 452 labour units over the 4-time period, whereas the labour demand reacts differently in the two countries.

The labour demand of the two health sectors declines about 19 labour units (country A) and only about 11 units in EU country B, as Fig. 16 shows.

In country A, labour demand in the other two sectors of the economy decreases by about 27%, while in country B, labour demand is reduced by only 7.2% for the FEW sector and 7.1% for the RoE sector.

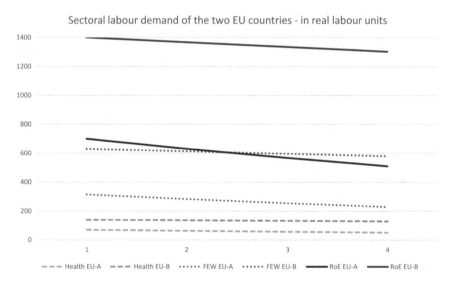

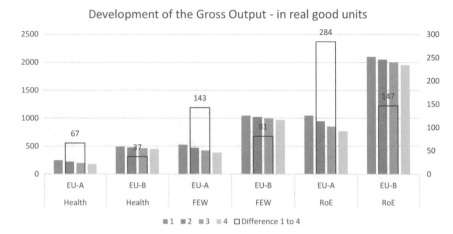

Fig. 15 *Source* Own calculations, 2021

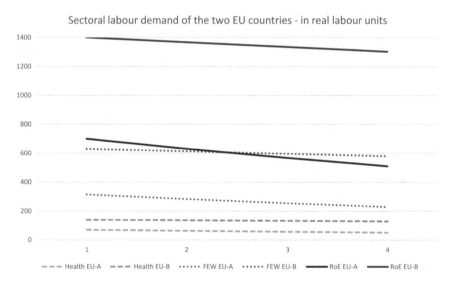

Fig. 16 *Source* Own calculations, 2021

6.6 Trade Relationships [T]

Figure 17 shows that the break-in of the two economies have also an impact on the trade relations of the two countries. Country B can generate a trade surplus from the second period on, so that country A has to face a small trade deficit of 0.04, which increases to −0.1 in the third and −0.2 in the last period. The economic break-in of

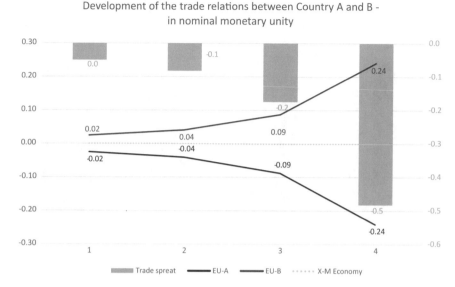

Fig. 17 *Source* Own calculations, 2021

the Corona pandemic leads to an increase of the trade deficit of country A and to a trade surplus of country B. The trade system (Export–Import (X-M)) of the two countries is over the 4-year period balanced. The export of the greater EU country benefits from the economic problems of the Corona pandemic.

So that, the trade spread between the two countries increases in the four-year period. The spread measures the absolut value (modulus) difference between the trade balance of the two countries. In period 1, the trade balance of both countries is balanced. However, with the assumed growth path, trade balances exhibit a surplus (country B) or deficit (country A) in each period. The difference increase to 0.1 in the second period and to 0.2 in the third and −0.5 in the last period.

6.7 CO_2-Emissions and Taxes [CO_2, CO_2 Tax]

Figure 18 shows the development of the CO_2 emissions of the two EU countries. The model results show a decrease in CO_2 emissions of about 27.1% in Country A and about 7.3% in Country B.

The development of the CO_2 emissions are influenced by the different CO_2 intensities (A = 0.23, B = 0.172) of the two countries.

If CO_2 emissions are now taxed then tax revenues show a similar distribution, see Fig. 19.

It can also be seen that over the observed period, the health sector only contributes between 16% in country A and 15% in country B to total tax revenues,

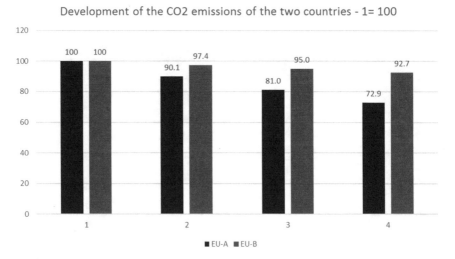

Fig. 18 *Source* Own calculations, 2021

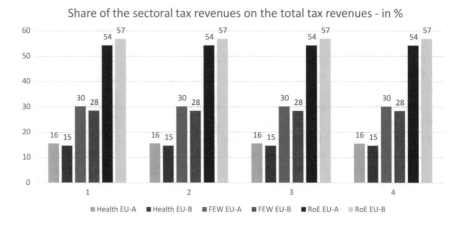

Fig. 19 *Source* Own calculations, 2021

followed by the FEW-nexus sector (30 and 28%) and the rest of the economy (RoE) sector. This sector generates between 54 and 57% of the tax revenues.

6.8 Utility and the Tax Compensation [Utility]

In the following, we analyze how the economic slump caused by the Corona measures affects the utility level of the households in the two countries. Utility measures the extent to which the needs of households are satisfied by the goods consumed. With each consumer good purchased, the household satisfies a piece of its need [86], i.e. the consumer registers in his consciousness the utility of these goods, so that utility can also be interpreted as the consumer's subjective evaluation of the commodities [86]. In the model economy, the ability to buy goods is limited due to declining income. To compensate possible losses of utility, the Kaldor-Hicks-Welfare-Criterion is taken up [43, 47, 86].

This welfare criteria is based on the idea of interpersonal compensation for changes in well-being caused by governmental measures such as taxes [86]. It belongs to the compensation criteria methods, such as the Scitovsky criterion or the criteria of Samuelson and Gorman [79]. The criteria attempts to offset welfare gains and losses against each other [51].

Figure 20 now shows that in the case where the government has not determined compensation method [no (sine)], the utility level of households in both countries decreases in the 4-period period. By 27% in the case of country A and by 26% for country B.

If the compensation concept is used by the government, then the utility decreases for country B only about 7% and in the case of country B, the utility level remains almost unchanged (-3.1%).

The Kaldor-Hicks compensation concept can support government policies toward a carbon-free economy as well as its Corona measures. The compensation concept can increase social acceptance and thus support sustainable development.

7 COVID-19 and Sustainability

In the current Corona time of uncertainty the German government published in 2021 its new Sustainability Strategy "Deutsche Nachhaltigkeitsstrategie – Weiterentwicklung 2021 – Kurzfassung" to present its concept for a sustainable development of Germany in a European context [32, 33] based on the 17 UN Sustainable Development Goals (SDGs) [95]. The German government stressed in its strategy that Germany is embedded in an international cooperation system and its new strategy will support the global community on the way to sustainable development [33]. In the view of the German government, the Corona pandemic has stressed the need for international cooperation to deal with the increased economic pressure caused by the Coronavirus and climate change [32].

Both challenges show that threats to one of the global sustainability goals—in the case of the corona pandemic to the health goal (SDG 3)—also create threats to other areas of social life and thus to their sustainability goals as well, as the following Fig. 21 shows. The Corona pandemic affects directly SDG 3 (health & well-being) and indirectly eight additional SDGs: SDG 4 (quality education), SDG 7 (clean energy), SDG 8 (decent work & economic growth), SDG 9 (innovation, infrastructure), SDG 10 (reduced inequalities), SDG 12 (responsible consumption & production), SDG 13 (climate action), and SDG 17 (fair trade).

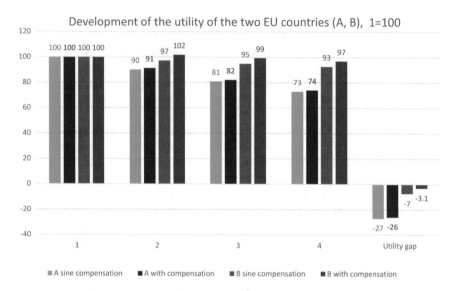

Fig. 20 *Source* Own calculations, 2021

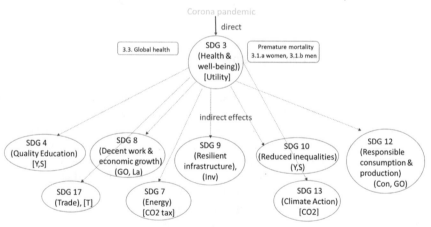

Fig. 21 *Source* Own calculations, 2021

The development of the nine SDGs can be measured by the indicators of the presented CGE model.

The Corona pandemic has a direct impact on SDG 3 "Good health and well-being". These effects can be measured by the indicators 3.1.a "premature mortality women", 3.1.b "premature mortality men", and 3.3 "global health". The indicator 3.3 measures Germany's contribution to the global pandemic prevention and response. This effect is captured in the CGE model by the development of the health sectors of the two EU countries and the well-being is measured by the utility indicator [utility] of the CGE model.

The Corona measures also affect the quality of education (SDG 4), as families face new tasks (home-schooling) and new expenses (notebooks for the children), while suffering declining incomes due to the Corona measures. These impacts are captured by the evolution of the income and savings indicators of the CGE model [Y,S].

Measures taken to mitigate COVID-19 can lead to economic disruptions and negative growth rates. These developments can jeopardize social opportunities for various social groups (SDG 8). To capture this effect, labour demand is analyzed in the CGE model through the labour indicator [La]. The increase in unemployment also leads to an increase in inequalities (SDG 10). The negative growth rates also affect the investments (SDG 9), which is measured by the investment indicator [Inv] of our CGE Model. Declining investments also affect the future perspective of responsible consumption and production behaviour (SDG 12). The development of SDG 12 is measured by the CGE model indicators consumption [Con] and gross output [GO].

The development of the model economy has also a negative impact on the trade relations of the two EU economies as the model results have shown. The development of the CGE indicator trade [T] measures the development of SDG 17 (fair trade). CO_2 emissions are briefly declining, as shown by the model indicator [CO_2], but the lack of investment may hinder the transformation toward a zero-emissions society (SDG 13). This development is measured by the CGE indicator [CO_2] emissions. The established CO_2 tax has an impact on the energy sector and on SDG 7 (modern energy). This impact can be measured by the CGE CO_2 tax indicator [CO_2 tax]. In summary, the impact of government actions to contain the Corona pandemic on sustainable development can be measured by the indicators of the CGE model.

8 Conclusion

The analysis has shown that the Corona pandemic is not the first health crisis to cause serious casualties worldwide, but it is the first global health crisis in which multiple vaccines have been developed within 12 months. Yet despite this success, the pandemic has cost millions of lives in just one year and caused severe socio-economic impacts around the globe.

The analysis has shown that the socio-economic effects of the pandemic and of the CO_2 debts crises can be analyzed by a CGE model. The model results have shown, that the gross output of the economies is reduced, the social inequality increased, the labour demand is reduced, the lack of investments in sustainable production and consumption patterns will affect the ability of the EU countries to organize the transformation of the countries towards sustainable development. The difference in the consumption of the health goods between countries A and B shows that the government must intervene to enable households to purchase health care goods even in difficult economic times.

The model results can support the national and regional discourse of societal actors to find solutions for sustainable development as outlined in the German sustainability strategy. The Kaldor-Hicks-Compensation-Criterion can be a resilience tool for restoring people's trust in national and international institutions.

The Corona pandemic has stressed the urgent need for international collaboration in international organizations such as WHO, the Covax Initiative and the Paris Agreement to address global challenges.

References

1. Abila DB, Dei-Tumi SD, Humura F, Aja GN (2020) We need to start thinking about promoting the demand, uptake and equitable distribution of COVID-19 vaccines NOW! Public Health Pract 1:100063. https://doi.org/10.1016/j.puhip.2020.100063
2. Amann M (2021) Danke für nichts vol 27.30.2021. Der Spiegel Verlag, Hamburg
3. Armington PS (1969) A theory of demand for products distinguished by place of production. Staff Pap (International Monetary Fund) 16:159–178
4. Arnold M, Fleming S (2020) Eurozone budget deficits rise almost tenfold to counter pandemic. FT, London
5. Asen E (2020) Carbon taxes in Europe. Tax foundation. https://taxfoundation.org/carbon-taxes-in-europe-2020/. Accessed 8 Oct 2020
6. Auswärtiges Amt der Bundesrepublik Deutschland (The German Federal Foreign Office) (2021) Corona-Impfstoff weltweit fair verteilen: GAVI kündigt erste Lieferung von Impfdosen über COVAX an. German government. https://www.auswaertiges-amt.de/de/aussenpolitik/themen/gesundheit/covax/2395748. Accessed 3 Apr 2021
7. Babatunde KA, Begum RA, Said FF (2017) Application of computable general equilibrium (CGE) to climate change mitigation policy: a systematic review. Renew Sustain Energy Rev 78:61–71. https://doi.org/10.1016/j.rser.2017.04.064
8. Bartels H (2019) Überblick über den Lastenausgleich. Paper presented at the Kriegsfolgenarchivgut, Bayreuth, 14.10.2019
9. BBC News (2021) Covid: South Africa halts AstraZeneca vaccine rollout over new variant. BBC. https://www.bbc.com/news/world-africa-55975052. Accessed 4 Apr 2021
10. Bergdolt K (2003) Der schwarze Tod - Die große Pest und das Ende des Mittelalters. C.H. Beck, Munich
11. Böhringer C, Rutherford TF, Wiegard W (2003) Computable general equilibrium analysis: opening a black box ZEW. Discussion paper
12. Bullard N (2010) Climate debt: a subversive political strategy. Clim Justice Now. https://www.tni.org/es/node/10897. Accessed 3 Mar 2021
13. Bundesamt für zentrale Dienste und offene Vermögensfragen (2021) Historie Lastenausgleich. Bundesamt für zentrale Dienste. https://www.badv.bund.de/DE/Lastenausgleich/Historie Lastenausgleich/Ausgangslage/start.html. Accessed 6 Mar 2021
14. Bundespräsident Frank-Walter Steinmeier (2021) Central act of remembrance for those who died in the pandemic, (published in German: Zentraler Gedenkakt für die Verstorbenen der Pandemie). Bundespräsidialamt. https://www.bundespraesident.de/SharedDocs/Reden/DE/Frank-Walter-Steinmeier/Reden/2021/04/210418-Corona-Gedenken.html. Accessed 19 Apr 2021

15. Cantor NF (1997) In the wake of the Plague—the black death and the world it made. Harper Perennial, London
16. Centers for Disease Control and Prevention (CDC) (2019) 1918 Pandemic (H1N1 virus). CDC. https://www.cdc.gov/flu/pandemic-resources/1918-pandemic-h1n1.html. Accessed 2 Feb 2021
17. Deutsche Welle (2020a) Corona-Krise: Japans Wirtschaft in der Rezession. Deutsche Welle,. https://www.dw.com/de/japans-wirtschaft-in-der-rezession/a-53475440. Accessed 18 May 2020
18. Deutsche Welle (2020b) COVID: EU to start vaccinations on December 27. Deutsche Welle. https://www.dw.com/en/covid-eu-to-start-vaccinations-on-december-27/a-55973609. Accessed 17 Dec 2020
19. Deutsche Welle (2020c) Viruserkrankung-Italien meldet erste Corona-Todesfälle. https://www.dw.com/de/italien-meldet-erste-corona-todesf%C3%A4lle/a-52473262. Accessed 23 Feb 2020
20. Dixon PB, Koopman RB, Rimmer MT (2013) Chapter 2—the MONASH style of computable general equilibrium modeling: a framework for practical policy analysis. In: Dixon PB, Jorgenson DW (eds) Handbook of computable general equilibrium modeling, vol 1. Elsevier, pp 23–103. https://doi.org/10.1016/B978-0-444-59568-3.00002-X
21. ECOMOD (2003) Practical general equilibrium modeling using GAMS. EcoMod Press, Northampton, MA
22. Edara VV et al (2021) Infection- and vaccine-induced antibody binding and neutralization of the B.1.351 SARS-CoV-2 variant. Cell Host Microbe. https://doi.org/10.1016/j.chom.2021.03.009
23. European Centre for Disease Prevention and Control (2021) COVID-19 situation update for the EU/EEA. https://www.ecdc.europa.eu/en/cases-2019-ncov-eueea. Accessed 7 Feb 2021
24. European Commission (2020a) Covid-19-Impfstoff für alle: Kommission unterstützt Initiative COVAX EU. https://ec.europa.eu/germany/news/20200831-covid-19-impfstoff-covax_de. Accessed 31 Aug 2020
25. European Commission (2020b) Recovery plan for Europe. European Union. https://ec.europa.eu/info/strategy/recovery-plan-europe_en. Accessed 21 July 2020
26. Eurostat (2021) BIP und Hauptkomponenten (Produktionswert, Ausgaben und Einkommen). Euorstat, Luxembourg
27. Forman R, Shah S, Jeurissen P, Jit M, Mossialos E (2021) COVID-19 vaccine challenges: what have we learned so far and what remains to be done? Health Policy. https://doi.org/10.1016/j.healthpol.2021.03.013
28. Frampton D et al (2021) Genomic characteristics and clinical effect of the emergent SARS-CoV-2 B.1.1.7 lineage in London, UK: a whole-genome sequencing and hospital-based cohort study. Lancet Infect Dis. https://doi.org/10.1016/S1473-3099(21)00170-5
29. Freund M (1979) Deutsche Geschichte. Bertelsmann Verlag, Munich
30. Fuhrer JC (1997) The (Un)importance of forward-looking behavior in price specifications. J Money Credit Banking 29:338–350. https://doi.org/10.2307/2953698
31. GAVI The Vaccine Alliance (2021) COVAX explained. GAVI. https://www.gavi.org/vaccineswork/covax-explained. Accessed 21 Feb 2021
32. German Federal Government (2021) Deutsche Nachhaltigkeitsstrategie - Weiterentwicklung 2021. Bundesregierung, Berlin
33. German Federal Government (2021) Deutsche Nachhaltigkeitsstrategie - Weiterentwicklung 2021 - Kurzfassung. Bundesregierung, Berlin
34. Ghebreyesus TA (2021) WHO Director-General's opening remarks at 148th session of the Executive Board. World Health Organization. https://www.who.int/director-general/speeches/detail/who-director-general-s-opening-remarks-at-148th-session-of-the-executive-board. Accessed 21 Feb 2021
35. Graham MS et al (2021) Changes in symptomatology, reinfection, and transmissibility associated with the SARS-CoV-2 variant B.1.1.7: an ecological study. Lancet Public Health. https://doi.org/10.1016/S2468-2667(21)00055-4

36. Grennes T, Caner M, Koehler-Geib F (2010) "Finding The Tipping Point—When Sovereign Debt Turns Bad". Policy Research Working Papers. https://doi.org/10.1596/1813-9450-5391
37. Guidry JPD et al. (2021) U.S. public support for COVID-19 vaccine donation to low- and middle-income countries during the COVID-19 pandemic Vaccine 39:2452–2457. https://doi.org/10.1016/j.vaccine.2021.03.027
38. Guo YM, Shi YR (2021) Impact of the VAT reduction policy on local fiscal pressure in China in light of the COVID-19 pandemic: a measurement based on a computable general equilibrium model. Econ Anal Policy 69:253–264. https://doi.org/10.1016/j.eap.2020.12.010
39. Guterres A (2021) Remarks to the security council open meeting on ensuring equitable access to COVID-19 vaccines in contexts affected by conflict and insecurity. United Nations. https://www.un.org/sg/en/content/sg/speeches/2021-02-17/ensuring-equitable-access-covid-19-vaccines-contexts-affected-conflict-and-insecurity-remarks-security-council. Accessed 21 Feb 2021
40. Haqiqi I, Horeh MB (2013) Macroeconomic impacts of export barriers in a dynamic CGE model. J Money Econ 8:117–150
41. Hauser R (2011) Zwei deutsche Lastenausgleiche: eine kritische Würdigung. Vierteljahrshefte zur Wirtschaftsforschung 80:103–122. https://doi.org/10.3790/vjh.80.4.103
42. Heilemann U, Rappen H (2014) Solidaritätszuschlag. Konrad Adenauer Stiftung (KAS). https://www.kas.de/de/web/soziale-marktwirtschaft/solidaritaetszuschlag. Accessed 7 Mar 2021
43. Hicks J (1939) The foundations of welfare economics. Econ J 49:696–712. https://doi.org/10.2307/2225023
44. Höhne N, Blum H, Skeie RB, Kurosawa A, Hu G, Lowe J, Gohar L, Matthews B, Nioac de Salles AC, Ellermann C (2011) Contributions of individual countries' emissions to climate change and their uncertainty. Clim Change 106:359–391
45. Hughes ML (2009) Shouldering the burdens of defeat: West Germany and the reconstruction of social justice. The University of North Carolina Press
46. IEA (International Energy Agency) (2019) CO_2-Emissionen in kg pro BIP-einheit. Paris
47. Kaldor N (1939) Welfare propositions in economics and interpersonal comparisons of utility. Econ J 49:549–552. https://doi.org/10.2307/2224835
48. Kant I (1785 (2012)) Groundwork of the metaphysic of morals (published in German: Grundlegung zur Metaphysik der Sitten)
49. Keogh-Brown MR, Jensen HT, Edmunds WJ, Smith RD (2020) The impact of Covid-19, associated behaviours and policies on the UK economy: a computable general equilibrium model. SSM-Population Health 12:100651. https://doi.org/10.1016/j.ssmph.2020.100651
50. Kim JH et al (2021) Operation warp speed: implications for global vaccine security The Lancet. Glob Health. https://doi.org/10.1016/S2214-109X(21)00140-6
51. Kleinewefers H (2008) Einführung in die Wohlfahrtsökonomie. Theorie - Anwendung - Kritik. Kohlhammer, Stuttgart
52. Konrad Adenauer Stiftung (2020) Lastenausgleich in der Corona-Krise Analysen & Argumente Juni 2020
53. Lu M (2020) Chart: debt-to-GDP continues to rise around the World. Visual Capitalist. https://www.visualcapitalist.com/debt-to-gdp-continues-to-rise-around-world/. Accessed 3 Mar 2021
54. Malpass D (2021) Building a green, resilient, and inclusive recovery: speech by World Bank Group President David Malpass at the London School of Economics. The World Bank. https://www.worldbank.org/en/news/speech/2021/03/29/building-a-green-resilient-and-inclusive-recovery-speech-by-world-bank-group-president-david-malpass. Accessed 29 Mar 2021
55. Matthews HD (2016) Quantifying historical carbon and climate debts among nations. Nat Clim Change 6:60–64. https://doi.org/10.1038/nclimate2774
56. Metcalf GE (2008) Designing a carbon tax to reduce U.S Greenhouse gas emissions. Rev Environ Econ Policy 3:63–83. https://doi.org/10.1093/reep/ren015

57. Metcalf GE (2018) Paying for pollution: why a carbon tax is good for America. Oxford University Press (in production), Oxford
58. Meyer LH, Roser D (2010) Climate justice and historical emissions. Crit Rev Int Soc Pol Phil 13:229–253. https://doi.org/10.1080/13698230903326349
59. NASA (2021a) 2020 tied for Warmest Year on Record. NASA. https://earthobservatory. nasa.gov/images/147794/2020-tied-for-warmest-year-on-record?src=eoa-iotd. Accessed 20 Jan 2021
60. NASA (2021b) Global temperature—global land-ocean temperature index. NASA's Goddard Institute for Space Studies (GISS). https://climate.nasa.gov/. Accessed 7 Mar 2021
61. Nature Climate Change Editorial (2018) How to pay the price for carbon. Nat Clim Change 8:647. https://doi.org/10.1038/s41558-018-0256-0
62. Platto S, Wang Y, Zhou J, Carafoli E (2020) History of the COVID-19 pandemic: origin, explosion, worldwide spreading. Biochem Biophys Res Commun. https://doi.org/10.1016/j. bbrc.2020.10.087
63. Posen A (2021) Making the most of their shot: the American rescue plan package. Intereconomics 56:127–128. https://doi.org/10.1007/s10272-021-0965-x
64. Pramuk J (2021) House passes $1.9 trillion Covid relief bill, sends it to Biden to sign. CNBC. https://www.cnbc.com/2021/03/10/stimulus-update-house-passes-1point9-trillion-covid-relief-bill-sends-to-biden.html. Accessed 10 Mar 2021
65. Radtke R (2021) Epidemien und Pandemien. Statista. https://de.statista.com/themen/131/ pandemien/. Accessed 1 Apr 2021
66. Ritchie H (2019) Who has contributed most to global CO_2 emissions. Our World in Data. https://ourworldindata.org/contributed-most-global-co2. Accessed 1 Oct 2019
67. Ritchie H et al (2021) Coronavirus (COVID-19) Vaccinations. Our World in Data. https:// ourworldindata.org/covid-vaccinations. Accessed 9 Feb 2021
68. rme/aerzteblatt.de (2021) SARS-CoV-2: Britische Variante hat Mutation E484K der südafrikanischen Variante übernommen. https://www.aerzteblatt.de/nachrichten/121049/ SARS-CoV-2-Britische-Variante-hat-Mutation-E484K-der-suedafrikanischen-Variante-uebernommen. Accessed 27 Mar 2021
69. Robert-Koch-Institut (Wolfgang Kiehl) (ed) (2015) RKI Fachwörterbuch Infektionsschutz und Infektionsepidemiologie. Robert Koch-Institut, Berlin
70. Robert Koch Institut (2021) Übersicht und Empfehlungen zu besorgniserregenden SARS-CoV-2-Virusvarianten (VOC). https://www.rki.de/DE/Content/InfAZ/N/Neuartiges_ Coronavirus/Virusvariante.html. Accessed 3 Apr 2021
71. Robert Koch Institut (RKI) (2021) Todesfälle mit Coronavirus (COVID-19) in Deutschland nach Alter und Geschlecht. Berlin
72. Robson EN, Wijayaratna KP, Dixit VV (2018) A review of computable general equilibrium models for transport and their applications in appraisal. Transp Res Part A: Policy Pract 116:31–53. https://doi.org/10.1016/j.tra.2018.06.003
73. Sadarangani M et al. (2021) Importance of COVID-19 vaccine efficacy in older age groups Vaccine 39:2020–2023. https://doi.org/10.1016/j.vaccine.2021.03.020
74. Saito M (2004) Armington elasticities in intermediate inputs trade: a problem in using multilateral trade data. Can J Econ/Revue canadienne d'Economique 37:1097–1117
75. Sanderson BM, O'Neill BC (2020) Assessing the costs of historical inaction on climate change. Sci Rep 10:9173. https://doi.org/10.1038/s41598-020-66275-4
76. Schnibben C (2021) Sind wir dümmer als dieses Ding ohne Hirn? Der Spiegel, Hamburg
77. Shen X et al (2021) SARS-CoV-2 variant B.1.1.7 is susceptible to neutralizing antibodies elicited by ancestral spike vaccines. Cell Host Microbe. https://doi.org/10.1016/j.chom.2021. 03.002
78. Shoven JB, Whalley J (1993) Applying general equilibrium. Cambridge University Press, Cambridge
79. Sohmen E (1992) Allokationstheorie und Wirtschaftspolitik. Mohr Siebeck, Tübingen

80. Statista (2021a) Anzahl der Todesfälle aufgrund ausgewählter Pandemien weltweit bis zum Jahr 2020. Statista. https://de.statista.com/statistik/daten/studie/1126584/umfrage/todesfaelle-aufgrund-von-ausbruechen-ausgewaehlter-infektionskrankheiten/. Accessed 20 Mar 2020
81. Statista (2021b) Epidemien und Pandemien 1918-2021. Statistisches Bundesamt, Wiesbaden
82. Statista (2021c) Geschätzte Entwicklung der Weltbevölkerung in den Jahren 10000 vor Christus bis zum Jahr 2000 Statista GmbH. https://de.statista.com/statistik/daten/studie/1066248/umfrage/geschaetzte-entwicklung-der-weltbevoelkerung/. Accessed 21 Mar 2021
83. Statista (2021d) Steuereinnahmen durch den Solidaritätszuschlag in Deutschland 2005 bis 2019. Statista. https://de.statista.com/statistik/daten/studie/30376/umfrage/steuereinnahmen-des-bundes-durch-den-solidaritaetszuschlag/. Accessed 3 Mar 2021
84. Statistisches Bundesamt (2021) EU-Monitor COVID-19. Statistisches Bundesamt. https://www.destatis.de/Europa/DE/Thema/COVID-19/_inhalt.html. Accessed 23 Feb 2021
85. Stern N (2006) The economics of climate change. The Stern review. Cambridge University Press, Cambridge
86. Stobbe A (1991) Micro economics [in German: Mikroökonomik]. Springer, Berlin
87. Taylor DB (2021) A timeline of the Coronavirus pandemic. The New York Times, New York
88. The Lancet (2021) Access to COVID-19 vaccines: looking beyond COVAX The Lancet 397:941. https://doi.org/10.1016/S0140-6736(21)00617-6
89. The White House (2021) President biden announces American rescue plan. The White House. https://www.whitehouse.gov/briefing-room/legislation/2021/01/20/president-biden-announces-american-rescue-plan/. Accessed 20 Jan 2021
90. Tobin J (1969) A general equilibrium approach to monetary theory. J Money Credit Banking 1:15–29
91. Toovey OTR, Harvey KN, Bird PW, Tang JW-TW-T (2021) Introduction of Brazilian SARS-CoV-2 484K.V2 related variants into the UK. J Infect. https://doi.org/10.1016/j.jinf.2021.01.025
92. UNAIDS (2020) Fact sheet—world aids day 2020 global hiv statistic. Unaids, Geneva
93. UNFCCC Paris agreement. In: Conference of the parties twenty-first session Paris, 30 November to 11 December 2015, Paris, 2015. UNFCCC
94. United Nations (2015a) Paris agreement. New York
95. United Nations (2015) Transforming our world: the agenda for sustainable development. United Nations, New York
96. United Nations Department of Economic and Social Affairs—Economic Analysis (2007) Capacity development: social accounting matrices data. UN DESA. https://www.un.org/development/desa/dpad/publication/capacity-development-social-accounting-matrices-data/. Accessed 29 Mar 2021
97. van de Ven P (2014) Social accounting matrix. In: Michalos AC (ed) Encyclopedia of quality of life and well-being research. Springer Netherlands, Dordrecht, pp 6010–6012. https://doi.org/10.1007/978-94-007-0753-5_789
98. Waffenschmidt H (1987) 35 Jahre Lastenausgleichsgesetz. Bundesregierung. https://www.bundesregierung.de/breg-de/service/bulletin/35-jahre-lastenausgleichsgesetz-806834. Accessed 24 Mar 2021
99. Wang Q, Han X (2021) Spillover effects of the United States economic slowdown induced by COVID-19 pandemic on energy, economy, and environment in other countries. Environ Res 196:110936. https://doi.org/10.1016/j.envres.2021.110936
100. White LA, Mordechai L (2020) Modeling the justinianic plague: comparing hypothesized transmission routes. PLOS (30 Apr 2020). https://doi.org/10.1371/journal.pone.0231256
101. World Health Organization (WHO) (2020a) Vaccines and immunization: what is vaccination? WHO. https://www.who.int/news-room/q-a-detail/vaccines-and-immunization-what-is-vaccination. Accessed 7 Jan 2021
102. World Health Organization (WHO) (2020b) WHO announces COVID-19 outbreak a pandemic. WHO. https://www.euro.who.int/en/health-topics/health-emergencies/coronavirus-

covid-19/news/news/2020/3/who-announces-covid-19-outbreak-a-pandemic. Accessed 1 Mar 2020

103. World Health Organization (WHO) (2021a) The ACT-Accelerator frequently asked questions. WHO. https://www.who.int/initiatives/act-accelerator/faq. Accessed 12 Jan 2021

104. World Health Organization (WHO) (2021b) Statement on the second meeting of the International Health Regulations (2005) Emergency Committee regarding the outbreak of novel coronavirus (2019-nCoV). WHO. https://www.who.int/news/item/30-01-2020-statement-on-the-second-meeting-of-the-international-health-regulations-(2005)-emergency-committee-regarding-the-outbreak-of-novel-coronavirus-(2019-ncov). Accessed 30 Jan 2021

105. World Health Organization (WHO) (2021c) WHO coronavirus (Covid-19) Dashboard. WHO. Accessed 10 Feb 2021

106. Wunderlich AC, Kohler A (2018) Using empirical Armington and demand elasticities in computable equilibrium models: an illustration with the CAPRI model. Econ Model 75:70–80. https://doi.org/10.1016/j.econmod.2018.06.006

107. Xu J, Zhang H, Başar T (2021) Stackelberg solution for a two-agent rational expectations model. Automatica 129:109601. https://doi.org/10.1016/j.automatica.2021.109601

108. Yadav PD et al (2021) Imported SARS-CoV-2 V501Y.V2 variant (B.1.351) detected in travelers from South Africa and Tanzania to India. Travel Med Infect Dis 41:102023. https://doi.org/10.1016/j.tmaid.2021.102023

109. Zhou D et al (2021) Evidence of escape of SARS-CoV-2 variant B.1.351 from natural and vaccine-induced sera. Cell. https://doi.org/10.1016/j.cell.2021.02.037

110. Ziano I, Li J, Tsun SM, Lei HC, Kamath AA, Cheng BL, Feldman G (2021) Revisiting "money illusion": Replication and extension of Shafir, Diamond, and Tversky (1997). J Econ Psychol 83:102349. https://doi.org/10.1016/j.joep.2020.102349

Holger Schlör studied economics at the University of Heidelberg and went on to complete his Ph.D. in economics in Berlin. He received a scholarship from the German Marshall Fund and the Alfried Krupp von Bohlen und Halbach Foundation. He has conducted research at several scientific institutions and the German Parliament. He is currently working at Forschungszentrum Jülich in the Institute of Energy and Climate Research—Systems Analysis and Technology Evaluation (IEK-STE). His research here focuses on the fields of sustainable development, economics and energy systems analysis. He was a member of the Scientific Committee for Social Sciences and Humanities of the Croatian Science Foundation. He was awarded the "Applied Energy 2017 Outstanding ICAE Paper" for his paper "The energy mineral society nexus—A social LCA model." He is the subject assistant editor of the Journal Applied Energy.

Stefanie Schubert is Professor of Economics at SRH University Heidelberg. Previously, she was assistant professor of organization theory and management at WHU-Otto Beisheim School of Management (Koblenz & Düsseldorf). Her expertise includes strategic decision making, managerial economics and strategic alliances and networks. In addition, Stefanie is consultant for strategic management with a particular focus on conceptual strategy development and strategic behaviour. She has published in distinguished international journals, such as the Journal of Health Economics, Applied Economics and Small Business Economics. She received her PhD from University Duisburg-Essen and graduated from Heidelberg University.

Waste Management of Medical Personal Protective Equipment and Facemasks: Challenges During and Post COVID-19 Pandemic

Unsanhame Mawkhlieng and Abhijit Majumdar

Abstract The ongoing COVID-19 pandemic has affected millions of people all over the world. To cope with this contagious disease, healthcare personnel use personal protective equipment (PPE) that includes gloves, face shields, goggles, gowns, coveralls, etc. Most of these items, if not all, are manufactured from non-biodegradable polymers like polypropylene, polyethylene, polyethylene terephthalate, etc. These plastics have a very long shelf life. The unprecedented and sudden increase in the use of facemasks and PPE has increased the amount of waste generated by several times. This has posed a challenge of handling both biomedical waste (BMW) and municipal solid waste (MSW). All of these when mismanaged and disposed discriminately find their way ultimately into the oceans. A detailed summary of the sustainability and waste management issues of facemasks and PPE is provided in this chapter. Present challenges of disposal, the short- and long-term effects of mismanagement, strategies and guidelines to assist proper disposal and possible immediate and futuristic remedies to alleviate the problem have been highlighted and deliberated.

Keywords COVID-19 · PPE · Masks · Sustainable · Recycle · Environment

1 Introduction

The ongoing COVID-19 pandemic that started at the end of 2019 has affected millions of people all over the world. The novel disease that began as a pneumonia of unknown cause in Wuhan, China was later found in January 2020 to be caused by SARS-CoV-2, a coronavirus that has its ecological origin in bats [1]. Shortly after its advent, cases of the new disease started emerging in other countries outside China, forcing the World Health Organization (WHO) to declare the outbreak a

U. Mawkhlieng · A. Majumdar (✉)
Department of Textile and Fibre Engineering, Indian Institute of Technology Delhi,
110016 New Delhi, India
e-mail: majumdar@textile.iitd.ac.in

© The Author(s), under exclusive license to Springer Nature Singapore Pte Ltd. 2021   37
S. S. Muthu (ed.), *COVID-19*, Environmental Footprints and Eco-design
of Products and Processes, https://doi.org/10.1007/978-981-16-3856-5_2

Public Health Emergency of International Concern on 30 January 2020 [2]. This calls for a strict comprehensive strategy of measures to contain and suppress the spread of the virus. As a step towards creating awareness and combating the spread of COVID-19, WHO has designed a webpage entitled "Coronavirus disease (COVID-19) advice for the public" that is dedicated to provide instructions about the precautionary steps needed to be taken, along with other relevant information [3]. Suggested steps include social distancing, avoiding crowded places, always wearing a mask in public places, keeping the surroundings sanitized, thorough washing of hands and coughing into a bent elbow or tissue. In places where the exposure to the disease is higher such as health centres, it is especially advised to remain hygienic and sterilized and to use personal protective equipment (PPE). More so, healthcare personnel caring for severe or critically ill COVID-19 patients are advised to suspend measures intended for rational use of PPE during shortages that include PPE extended use and reprocessing followed by reuse [4]. Thus, it is understood that the use of masks by the public and single-use PPE by health workers has increased substantially. The proportion of increase in mask usage takes precedence because billions of people wear a mask or two a day and it is the most used component of PPE around the world presently. In fact, according to Ocean Conservancy, a staggering 129 billion face masks are estimated to have been used every month of this pandemic [5]. To put this in perspective, the area of all manufactured masks when sewn together is large enough to cover the landmass of Switzerland [6]. Such a colossal consumption of masks is naturally associated with the problems of disposal. Assuming that one mask weighs an average of 4 g, then where does this 5500 tons of discarded masks end up? [7] The answer is "the ocean". Opération Mer Propre, a French NGO that works on the conservation of oceans captured some chilling images showing how masks and gloves float freely in the water bodies. Two of the images are shown in Fig. 1 [8]. According to other similar organizations working for the conservation of oceans and beaches such as RE-THINK and Oceans Asia, the discarded masks become a part of the myriad of marine debris [7, 9]. And this is just the masks used by the public!

Fig. 1 Masks and PPE pollutants of the ocean [8]

Healthcare personnel use PPE gear that includes other components as well such as gloves, face shields, goggles, gowns, coveralls, etc. and most, if not all, are single used. Surgical and respiratory masks are generally made of polypropylene whereas gloves, of materials such as polyvinyl chloride (PVC) or nitrile butadiene rubber (NBR). Disposable isolation gowns are often manufactured from nonwovens of polypropylene, polyethylene and polyester while face shields and goggles are products of either polycarbonates or polyethylene terephthalates. All these materials are non-biodegradable and have a very long shelf life. WHO estimated that the global need of these monthly supplies for the front line health workers to protect themselves and others from COVID-19 stands at an alarming 89 million masks, 76 million gloves, 30 million gowns and 1.6 million goggles [10].

A summary of the issues related to the sustainability and waste management of masks and medical PPEs are provided in this chapter with the hope to combat this emerging problem strategically. The chapter is aligned to highlight the present challenges of disposal; the short- and long-term effects of mismanagement; approaches to reduce, reuse and recycle masks and PPE; and possible immediate and futuristic remedies to alleviate the problem.

2 Types of Masks and Respirators

Based on functionality, masks can be classified under three categories: (1) filtering respirators/filter masks, (2) surgical/medical masks and (3) non-medical/cloth mask.

2.1 Filtering Respirators

Filtering facepiece respirators (FFPs) are also simply referred to as filtering respirators or filter masks. As the name suggests, respirators are intended to filter contaminants from the air that is breathed in, ensuring that there is minimal ingress of hazardous particulates and microorganisms into the wearer's body. Respirators are varied and are designed based on the level of protection they provide. For example, an N95 respirator is different from a P100 respirator. Respirators can also be classified as full masks or half masks depending on the protected area of coverage. The former guards the entire face, covering the eyes, mouth and nose whereas, the latter protects only the mouth and nose. Figure 2a–c show different types of face masks or respirators that are manufactured following certain sets of standards. Depending on the level of leakage of particulate matter into the interior of the respirators, the European standard, EN 149:2001-A1:2010 standard establishes 3 levels of protection and marked these respirators as FFP1, FFP2 and FFP3 having an inward leakage of 22%, 8% and 2%, respectively. On the other hand, the North American standard 42 CFR Part 84 developed by the National Institute for Occupational Safety and Health (NIOSH) establishes nine types of filters, namely,

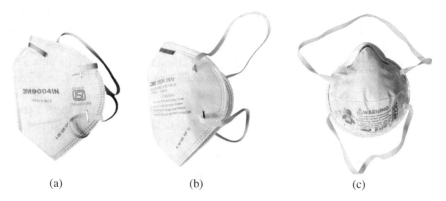

Fig. 2 Types of mask: **a** FFP1; **b** FFP2; **c** NIOSH N95

N95, N99, N100, R95, R99, R100, P95, P99 and P100 [11]. The letters N, R and P indicate the type of masks that can be used in the presence or absence of oil. N (not resistant) masks can be used when there is no oil in the air. Both R (oil resistant) and P (oil proof) type masks can be used when there are oil particulates in air. The difference between the two is that R type masks can only be used once for a span of not more than 8 h whereas P type masks can also be used multiple times according to the manufacturer's limit of reuse. The numbers 95, 99 and 100 indicate the filtration capacity that is determined against aerosols having 0.3 μm as mass median aerodynamic diameter.

2.2 Surgical Masks

Medical masks or surgical masks are medical devices, intended to cover the half face (nose, mouth and chin) to build a barrier, thereby limiting or preventing the transition of infectious agent/microorganisms or contaminants, body fluids and particulates between healthcare professionals or workers and patients [12, 13]. A typical surgical mask is shown in Fig. 3a. However, WHO does not recommend medical face masks at places with aerosol generating procedures [14].

2.3 Non-medical Cloth Masks

Non-medical masks, commonly known as cloth/fabric masks, home-made masks, etc., are majorly intended for the class of population with no symptoms or for those who are not exposed to high-risk environments. This category of masks includes various types of home-made or commercially available face masks developed from a single or multiple layers of a regular fabric (Fig. 3b). Also, termed as community

(a)   (b)

Fig. 3 **a** Surgical mask; **b** Cloth mask

masks by European Centre for Disease Prevention and Control (ECDC) and cloth face coverings by the Centers for Disease Control and Prevention (CDC), these masks are strictly not intended for use in healthcare settings or by any healthcare professionals or workers.

2.4 Filtration Performance

The effectiveness of masks is undoubtedly dependent on their proper usage [15]. However, each mask type strongly differs in their filtration capacity. For instance, on comparing the effectiveness of cloth with medical masks in a hospital scenario, MacIntyre et al. reported that the penetration of particles through the medical masks was 44%, whereas, for a cloth mask, it was almost 97% [16]. This indicates that these kinds of improvised home-made cloth masks should be used only at places with low risks as they tend to increase risks of infection due to humidity, virus retention or liquid diffusion. Hence, for airborne infective agents or particulates with particles sizes of 0.022–0.259 µm, FFPs (such as N95) that show a maximum filtering efficiency of 95% are recommended by CDC. Therefore, respirators such as N95 FFP can provide the desired respiratory protection to the wearer since the aerosol droplets containing the virus are usually greater in size than the 0.120 µm virus itself [14]. Research studies conducted to compare the performance of N95 respirators to surgical masks did not find significant evidence to support the superiority of the former against bacteria and viruses or acute respiratory infections in a clinical setting although, in a laboratory setting, the N95 respirators seemed to be more effective [17]. Additionally, Radonovich et al. reported no real difference on comparing the performance of N95 respirators to that of surgical masks against influenza and other related respiratory infections in laboratory settings [18].

Although the filtration efficiency of cloth masks is quite low, the severity of mask shortage and PPE supply disturbance have compelled CDC to recommend the use of cloth mask for the general public and non-healthcare professionals at low-risk areas to cover the face [14]. One of the challenges of cloth masks is that they vary widely in their performance since different kinds of constructions and materials are used. With this understanding in mind, Zhao et al. studied the filtration properties of different masks that are likely to be used by the general public by employing a modified procedure that is usually adopted for the approval of N95 respirators [14]. The masks chosen for the study were made of natural and synthetic materials. It was reported that commonly used woven and knitted fabrics such as cotton, polyester, nylon and silk give a filtration efficiency of 5–25%. Other items that are usually not a choice for mask construction such as tissue and copy paper showed good filtration efficiency. The authors stated that the filtration efficiency of polypropylene spun bond can be enhanced (from 6 to >10%) by charging the surface triboelectrically.

The need for masks and PPE is undisputedly necessary but the associated environmental concerns cannot be overlooked. Therefore, a drive towards reusing, recycling and wearing masks and PPE made from organic and biodegradable materials should be a top priority. Existing alternatives should be considered and new approaches that compete as closely as possible to the prevalent current materials both in terms of performance and cost should be encouraged.

3   Present Challenges of Waste Management

Plastic generation and usage have been under check for many years, with certain regulations imposing some form of ban on single-use items and bags. However, the increased use in single-use plastics during the COVID-19 pandemic time may change the course of plastic use policies around the world for good. The unprecedented and sudden increase in masks and PPE has increased the amount of waste generated several folds and that poses a challenge of handling both biomedical waste (BMW) and municipal solid waste (MSW). The flow rate of the two business as usual (BaU) wastes over time is shown in Fig. 4. While the flow rate of MSW has slightly reduced during the peak time frame, the flow rate of BMW has increased multi-fold.

As far as the infectious BMW is concerned, the increase in the amount during the pandemic is mind-boggling. For instance, in Hubei Province of the China alone,

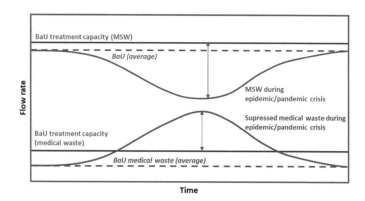

Fig. 4 Flow rate of BMW and MSW as a function of time during the pandemic crisis [19]

an increase of 500% from 40 tons per day to 240 tons per day was observed during the peak of the pandemic [20]. The heap of waste generated is to be supposedly handled by the present waste management system that is barely able to tackle the existing pre-pandemic waste efficiently, especially in developing countries. As an example, an article published on The Print on 5th September 2020 reveals jaw-dropping facts about India's BMW management that was unable to treat 600 tons of waste per day even before the pandemic [21]. The upsurge in the amount of BMW stems from two reasons; firstly, the inherent increased used of PPE and secondly, household waste is now treated as BMW due to home quarantining and self-care of asymptomatic patients. The situation further burdens the waste management systems in countries like India that have a disproportionately miniscule number of waste treatment centres and health care facilities (HCFs) with in-house captives. For example, India has only 198 BMW Treatment Centers (BMWTCs) as of 2018 [22] and merely 16% of HCFs, as of or on 2013, have their own disposal facilities [23]—figures inordinate to the country's population (1.38 billion, 2020). The end destination of these BMW is the landfills. Although the problem is universal, poorer nations suffer the most as they cannot afford to install sufficient treatment facilities in proportion to the waste generated. A similar challenge is evident in the case of MSW management as well, where above all else, the safety of the sanitary workers and garbage pickers is at risk.

The problem associated with waste management originates at the stages of sorting and collection, which is more aggravated in countries with lenient laws. Sometimes, the answer does not lie in 'more incinerators' as opinionated by experts who have expressed their concerns with the failure of waste segregation at the source [24]. Non-segregation of waste creates burden on the treatment plants which are designed to handle specific waste type. Worse, mixing of waste leads to incineration of contaminated items that should have been sterilized and treated before disposal, adding to the hazard. COVID-19 has taken the problem to a different level since general municipal household wastes are now mixed with potential BMW, given that many patients are under home treatment and containment zones are on the rise.

Another challenge that is now faced by the waste management system, particularly with developing countries is the lack of funding that is urgently needed to overcome the present scenario of waste rise. For instance, the European Commission has adopted the Coronavirus Response Investment Initiative where cohesion policy funds are readily available to Member States' budgets, allowing them to reallocate their finances and address waste management problems [25]. The same level of financial aid is not available in other developing countries. In fact, many do not even have the financial standing to install temporary mobile incinerators, a short-term approach to overcome the issue, as did Wuhan, for example [26].

The current pandemic has also exposed the unpreparedness of the present waste management system to emergencies, suggesting that the challenge could have been mitigated had the current system prepared itself better against unanticipated disasters.

3.1   Stages of Biomedical Waste Disposal

The disposal of BMW can be categorized into four stages: collection and segregation, storing and transporting, treatment and disposal [27].

3.1.1   Collection and Segregation

The BMW should be collected in strong and resilient containers, dumpers, or compactors to prevent them from leakage during the handling process. Since BMW is a mixture, it is very important to properly segregate them to prevent secondary infections. For example, used needles, syringes, sharps like blades or other contaminated tools should not be placed in common recycle bins or waste disposal. Similarly, liquid and solid wastes should be disposed separately and each type of segregation should be done in colour coded and labelled containers or bags.

3.1.2   Storing and Transporting

A secured area with specific requirements or facilities that is inaccessible to the public and separated from the area of food consumption should be assigned as a storage space for the waste products to prevent transmission of infections. Some storage facilities may provide special transportation and protective devices to handle, transport or dispose, the BMW products.

3.1.3   Treatment

The treatment of BMW products can broadly be categorized into two methods: incineration and non-incineration.

Incineration (Type 1 medical waste treatment): This method employs a high temperature thermal process to convert the BMW into gas, ash and heat through the combustion process. There are basically three kinds of incinerators that are used worldwide:

- The multiple hearth type incinerator that is a circular steel furnace housing having several hearths constructed vertically one above the other. The waste passes through the rotating hearths whose temperature gradually reduces from the top to the bottom of the incinerator. The burnt waste is collected in the form of ash at the bottom whereas hot gasses flow upward to the exhaust. There is a central cooling shaft that maintains the temperature of the furnace.
- Rotary kiln incinerator that consists of two cylindrical tanks, one of which is inclined and the other is placed vertically. The rotation of the inclined drum facilitates the waste movement towards the ash collector. It is in this chamber

that the waste is converted to gases through partial combustion, volatilisation and destructive distillation. The volatile gases generated pass to the vertical secondary drum where the completion of the combustion of organics in the flue gas takes place.

- Controlled air incinerator that contains two process chambers to handle the BMW in two stages. Firstly, the medical waste is fed to the primary chamber which is characterized by low air supply for incomplete combustion. The combustion gases then pass into the upper chamber where complete combustion takes place.

Incineration is a very effective method of treating hazardous and contaminant medical wastes with a high-volume reduction of around 80–90% [28]. Further, incineration is a waste-to-energy technology, leading to energy generation that can be efficiently used for electricity or heat generation. There is no generation of methane gas and there is minimal contamination of the underground water. However, incinerators are expensive and increase air pollution. It is also known that there are long-term health issues associated with incineration.

As far as MSW is concerned, incineration is not a recommended solution in Asia for two main reasons—improper waste segregation and high installation cost. Incinerators are limited to what they can burn. For effectiveness, low moisture and organic content are favourable. In Asia, the general practice is to segregate the waste at the landfill after collection. Hence, waste contamination is inevitable and effective waste segregation is impossible. For example, plastic materials are soiled with vegetable remains and vice versa. The organic components of the MSW which can be best dealt with decomposition often contain high moisture content and their burning affects the performance of the incinerators. Another problem with ineffective waste segregation is the amount of waste that has to be sent for incineration which often exceeds the treatment capacity [29].

Non-incineration (Type 2 medical waste treatment): The non-incineration treatment contains four basic processes, namely thermal, irradiation, chemical and biological. Thermal or autoclaving system requires high temperature that produces steam which decontaminates the BMW products. Autoclaving is commonly used for human body fluid waste, microbiology laboratory waste and sharps. However, this method is not applicable for cytotoxic agents used during treatment like chemotherapy as the generated waste are not degraded with autoclave steams. Irradiation is also a thermal method that uses a high frequency microwave to dispose the BMW. Briefly, the high frequency wave generates heat to the BMW products that in turn kills bacteria, or any other contaminations. Chemical method or decontamination is used to treat human blood and body fluid waste, sharps and microbiology laboratory waste. However, this method cannot be used to treat anatomical waste. Biological method or decontamination uses enzymes to destroy the organic matter of the waste; however, only few non-incineration ways of BMW treatment are based on biological methods.

In general, non-incineration technologies are safe and efficient. In fact, a study initiated by the National Institute of Occupational Safety and Health (NIOSH) revealed that the amount of volatile compounds which are organic in nature did not exceed that permissible limit prescribed by the Occupational Safety and Health Administration (OSHA) [30]. Additionally, the level of metal samples in the air was also found to be minimal, below the limits of detection. However, the main disadvantage of non-incineration is that the ergonomics of the facilities are usually not favourable as much of the activities involved in handling and emptying heavy waste bins are manually carried out. Such practices pose serious health risks to workers who are exposed to hazardous blood splatters and physical injuries. Consequently, the inherent safety issue inevitably compels the extensive use of PPE.

3.1.4 Disposal

Once the BMW are treated, the next step is to find the best way to appropriately dispose them by adhering to the regulations and guidelines. For example, in the US, municipal landfill and sanitary sewer systems may be used as final disposal places for the treated and decontaminated solid BMW. However, for fluids waste, every state and local government has their own rules, regulations and guidelines to properly dispose them. In general, the two recommended ways to handle biomedical fluid waste are:

- Fluid waste is collected in a leak proof container and solidified to further undergo autoclaving
- Autoclaved fluid wastes are then disposed into the sanitary sewer system

Nevertheless, precautions must be taken carefully prior to disposing the treated biomedical fluid waste in sewers as they may clog and leak.

4 Effects of Waste Mismanagement: Short- and Long-Term

The effect of waste mismanagement during the pandemic has diabolical consequences, some of which are expected to span for a long time. Hence, both the short- and long-term repercussions cannot be overlooked and are needed to be analyzed so that plausible remedies can be suggested.

4.1 Short-Term Effects

It is reported that improper waste management during COVID-19 can only escalate the risk of infection particularly among unprotected and unaware cleaners [31].

SARS-CoV-2, known to be long-lived, up to 72 h on plastic and steel surfaces, can be transmitted while handling the biohazard waste [32]. For instance, in Pune, India, waste mismanagement, coupled with poor safety measures, had led to an infection of over 100 families of waste-pickers [33]. Experts have also warned the potential domino-effect of COVID-19 waste to spread the virus through the waste-pickers. Given that most of them live in slums and crowded localities, the chances of infection to their families and others around is sky-high [34].

Another serious problem of waste mismanagement and the use of single-use disposable masks is the overflowing of drains and clogging of sewers. Masks that are indiscriminately thrown in public places either cause obstruction to the flow of water in drain or worse, pass through the drain grate and ultimately find their way into the ocean, a serious concern as mentioned earlier with long-term impact. Drainage overflow, leading to stagnant pools, is known to have caused several diseases, particularly in slums and overcrowded neighbourhoods. In fact, COVID-19 itself has a potential of spreading through such drain waters since evidence of SARS-CoV-2 in wastewater has been detected. In fact, Hart and Halden reported that the SARS-CoV-2 load in municipal wastewater lies in between 56.6 and 11.3 billion viral genomes per infected person per day [12]. In addition, cases of sewer clogging are of major concern, especially in the US, since it has been observed that citizens flush wipes, tissues, face masks and even rubber gloves down the toilet. A similar health hazard is associated with sewage water coming from toilets and sewers since there is a strong possibility of faecal–oral transmission in COVID-19 infection after it was found that faecal samples and anal swabs showed presence of the contagious viral nucleic acids [35, 36]. The ultimate consequence circles back to increased infection.

4.2 Long-Term Effects

The long-term effects of the pandemic waste mismanagement are already observed in beaches and oceans. The increase in the usage of single-use plastic items has already been discussed. To add to the problem, there is a reduction in recycling of recyclable waste due to fear of contagiousness. For instance, in the UK and US alone, around 46% and 31%, respectively, of the material recovery facilities have either stopped or reduced their operations [37]. The obvious implications are overburdened landfills or incinerations. However, mismanagement causes the non-recycled plastic items to reach the environment. It is estimated that the filter of a mask, made of polypropylene, will take around 500 years to decompose [38]. Disposable gloves take variable amounts of time to decompose depending on the material used. Natural latex decomposes in five years, whereas vinyl and nitrile gloves take decades, if not hundreds of years to decompose [38]. Hence, it is daunting to imagine if 129 billion face masks and 65 billion gloves used globally every month end up in the ocean [5]. A vicious threat is, thus, posed on marine life. Entanglement in masks and PPEs leads to injury and mortality of sea animals.

Fishes have also been known to ingest plastic, inducing a sense of fullness and reducing stomach storage, thereby causing starvation [39]. Many instances of dead seabirds with stomach full of plastic have been found even before the pandemic. A paper published in 2015 estimated that a concerning 99% seabird species would ingest plastic by 2050 and that effective waste management can be helpful to stop this from happening [40]. Therefore, it would not be surprising if this staggering figure of seabirds ingesting plastic reach sooner than 2050 should the present waste management not improve. Furthermore, a recent paper in 2020 revealed that the odour from marine debris due to algae and fungal growth on the floating plastics is often confused as food by sea turtles [41].

Additionally, waste that was accumulated during the pandemic, piled onto the landfills is expected to take many years to level since the rate of waste inflow is tremendous, especially in the pandemic times. An article published in September 2020 in the Times of India reported that it may take up to 15 years to process the current waste in Ghazipur, India alone [42]. The incoming waste in the near future is yet to be thought of. It is also anticipated that the water percolation and soil serration would also be negatively affected. There is also an expectation of increased energy consumption to deal with the present issue in the years to come.

5 Approaches to Reduce, Reuse and Recycle Masks and PPE

While focusing on better waste management system is indispensable, a more sensible approach is to limit the generation of waste. The process will not only have a positive impact on the financial strain of the economy, but also lead to a more sustainable environment. Less is the production of new plastic masks and PPE, lesser would be the pressure on waste segregation and better would be the waste management. Thus, in view of sustainability, this section is dedicated to various suggestive strategies to either reduce or reuse and recycle masks and PPE.

5.1 Use of Washable and Reusable Cloth Mask

The CDC, US suggests cloth masks for people who may have the virus and unknowingly transmitting it to others [43]. The use of cloth masks by the public is still debatable, however, it is generally accepted that it is good enough to slow down the spread of the virus. To comprehend this, it is crucial to know that the virus is contained in droplets and not as individual entities and thus, preventing the droplets from being inhaled will fence the transmission. The intention of wearing a mask is not to prevent inhaling droplets containing the virus, it is rather a strategy to contain the virus-rich droplets, from an infected person, within the mask. Thus, the slogan 'my mask protects you, your mask protects me' is campaigned worldwide. However, a recent unproven, although biologically plausible hypothesis of "inoculum", advocates for the productivity of the mask to the wearer himself. In essence, the hypothesis states that exposure to SARS-CoV-2 in unhazardous mild quantity, due to partial filtration by the mask, could lead to greater community-level immunity [44, 45]. However, until proven, health workers, elders who are over 50 years, people with underlying health conditions and others who are at the front line should use medical masks at all times [46].

The application of reusable and washable cloth mask reduces the severity of plastic production and waste. Although the three-layered mask recommended by WHO is supposed to contain a middle layer of spun-bonded polypropylene, however, the major portion of it is a woven fabric of cotton and is washable (Fig. 5). With each mask containing a lesser proportion of plastic and reusable a number of times, the level of plastic generation is expected to lower [47]. Where such masks are unavailable, double layered cloth masks of fine yarns and high thread density of over 400 threads per inch are expected to provide suitable filtration of around 90% [48]. Washing of masks at 60 °C with laundry detergent daily is recommended [49]. The number of washing cycles can be extended if the mask is not deformed and not worn out. Thus, the use of surgical masks may be restricted to a section of society, whereas for the public, reusable masks of woven fabric construction can be safely used with proper precautions.

5.2 Reuse of Masks After Decontamination

Earlier the usage of masks was dedicated to restricted work area or in medical scenario only, however, with COVID-19 creating a worldwide pandemic, it has

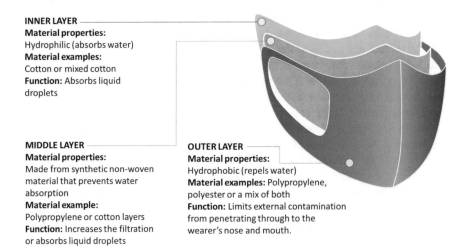

INNER LAYER
Material properties:
Hydrophilic (absorbs water)
Material examples:
Cotton or mixed cotton
Function: Absorbs liquid droplets

MIDDLE LAYER
Material properties:
Made from synthetic non-woven material that prevents water absorption
Material example:
Polypropylene or cotton layers
Function: Increases the filtration or absorbs liquid droplets

OUTER LAYER
Material properties:
Hydrophobic (repels water)
Material examples: Polypropylene, polyester or a mix of both
Function: Limits external contamination from penetrating through to the wearer's nose and mouth.

Fig. 5 Cloth mask recommended by WHO

become necessary to protect everyone. It is a general recommendation that medical face masks, FFPs and other PPE be used only once as originally intended. However, due to shortage of supply, reuse of masks and PPE has become important, compelling the research fraternity to delve into the possibilities of disinfection and decontamination of used equipment [50]. For reuse purposes, it is vital to know the longevity of the SARS-CoV-2 virus on inanimate surfaces. Studies have shown that the coronavirus can last up to a period of 72 h on surfaces [51] and therefore, it was suggested by the CDC that the masks and PPE can be reused after isolating them for at least five days [11]. This implies that everyone should possess at least five equipments, which may not be readily available during the pandemic. Therefore, disinfection or decontamination is bound to be adopted as alternative approaches.

Decontamination can be done either through chemical (involving the use of chemicals) or physical means (involving the use of air, radiation and UV rays). One of the physical methods that have been explored is heat inactivation. For instance, Wuhui et al. [52] reported that the efficacy of heat treatment on the influenza virus by using a hot dryer. They observed that exposure to dry hot air for 30 min could inactivate the virus effectively, whereas only partial inactivation was observed when oven baked at 56 °C for 30 min. The filtration efficiency was observed to be insignificantly affected. Thus, the authors suggested this technique of dry air as a quick and short-term immediate remedy where shortage of FFPs is severe. Autoclaving is another effective way to disinfect the used FFPs. Reported data demonstrate feasible reuse of FFPs that pass fit testing when autoclaved once and 86% passing when autoclaved for the second cycle [53]. Although with every autoclave cycle, FFPs substantially reduce its fit, however, stocks increase by around 66% [53].

Similarly, custom-built ultraviolet (UV) germicidal irradiation system, that can be developed using readily available (UV) lamp, is a cost-effective way to quickly decontaminate PPE or filtering respirators on site [54]. However, standard UV safety procedures must be considered rigidly to avoid direct exposure of eye or skin to the UV light source. Additionally, UV treatment is known to irradiate only the exposed surface and not the inside of the respirator due to shadow effects associated with multi-layered design of masks. Thus, in relation to UV treatment, UV-C is deemed more appropriate as it has more photon energy and can transmit through the FFP as well [55]. Decontamination by ozone has also been explored as another plausible approach. Zhang et al. [56] and Dennis et al. [57] reported that ozone can be used as an improvised solution to emergency PPE shortage since they found that the virucidal (deactivation of viruses) action was found to be faster than PPE degradation. Another advantage of ozone treatment over UV radiation is that the former can reach the crevices and shadowed areas of the PPE.

Further, the use of FDA approved vaporized hydrogen peroxide (VHP) is another promising approach to disinfect FFPs, enabling the health workers to retain their own respirators [58]. Although both UV systems and VHP demonstrate minimal to no impact on filtration with significant reduction in viral loads, the former demands an ample amount of time to ensure adequate exposure to UV light for disinfection and may occasionally encounter partial disinfection of straps. Besides, the scientific technical committees do not recommend the disinfection and sterilisation of disposable mask for reuse as it may be susceptible to other contaminations and possible reduction in filtering capacity. With documented reports suggesting the survivability of the virus from 48 h to 9 days [59], disinfecting treatments demand extensive evaluation to validate that the virus no longer persists in the masks. Hence, it is necessary to look for an alternative to the reuse of synthetic masks, which in turn would limit the impact on the environment by decreasing the rate of disposal either for landfilling or for incineration.

To combat these possible adversities, recycling of mask material is another alternative and innovative approach. One such evidential report examined and validated material recycling protocols using the most diffused type of disposable masks [60]. The team proposed different mechanical recycling protocols for each material composition of disposable masks leading to a recycling index ranging from 78 to 91%. In this context, recycling index refers to the amount of material that can be converted to a recycled raw material that possesses the required properties for mask construction. For example, in one of the strategies, the authors processed three layers of a facemask by melt extrusion and found that 78% of the mask weight was recycled. Hence, the recycling index was 78%. Further, when the ear loop pieces of the masks were also included in the recycling strategy, the index increased to 91%. Similarly, graphene functionalised masks with self-cleaning and photothermal properties may be explored as a unique, innovative and economical way to reuse or recycle the commercially available nonwoven surgical masks [61]. The graphene coated mask undergoes quick elevation in its surface temperature (over 80 °C) when illuminated under sunlight leading to self-sterilisation or self-cleaning for reuse. In addition, the graphene functionalised temperature sensitive masks provide a super hydrophobic surface that protects against the incoming respiratory droplets. Moreover, the presence of microporers within the masks promotes better salt-rejection performance for extended usage.

5.3 Reusable Elastomeric Respirators

Given the persistent worldwide shortage of PPE, the CDC, US recommended the conservation of disposable N95 masks [62]. However, disposable N95 masks are not designed to be reused and this causes significant safety concerns to the healthcare professionals [63–65]. In addition, with evidences demonstrating the presence of higher concentrations of aerosolised COVID-19 viral particles in rooms where PPE is doffed [63, 66] and its existence in the air for 3 h after aerosolisation [51], it is logical for healthcare workers and those at the front line to opt for PPE with full aerosol protection. In this regard, use of reusable elastomeric respirators with efficiency $\geq 95\%$ may replace hundreds to potentially thousands of disposable N95 masks. The National Academies of Science, Engineering and Medicine describes the kinds of air purifying respirators that are presently used in healthcare

[67]. In contrast to N95 masks, elastomeric respirators contain separate exhale vents and thus, the exhaled air does not pass through contaminated filters and trapped viral particles [63]. It is estimated that respirator masks with particulate filters can be used for at least one year in a hospital setting as long as the filter is not soiled or damaged [68]. In addition, elastomeric respirators provide 60% higher filtration performance and better seal as compared to the disposable respiratory masks having similar filter efficiency [51, 69]. Although less effective as compared to powered air purifying respirators, considering the severe shortages of PPE and considerable infection risk, the use of reusable elastomeric respirators would not only contribute to the safety of the PPE wearer, but they would also act as a step towards environmental cleaning and economical solution. In addition, with an elastomeric respirator, one can easily clean or wipe the external surface while it is still on as opposed to N95 which when put on for extended hours require additional use of a disposable surgical mask or face shield [63]. However, the wearer of elastomeric respirators must be cautious of any signs of active respiratory infections in themselves as ignoring this may increase the risk of contaminations and spreading infections. Moreover, special attention must be given to the use, storage and reuse of the elastomeric respirators particularly during donning and doffings of respirators throughout the clinical period so as to prevent contamination at the inside of respirators that would otherwise likely increase the risk of infection transmission within and outside healthcare settings [63].

6 Immediate Action and Future Recommendations

To overcome the current problems of mask and PPE related plastic pollution, a tight collaboration and cooperation among the natural and social scientists, the policymakers, the industrialists, the waste managers and the public is of utmost importance. The present scenario has opened new opportunities to renew policies and to realign the way of living to address the problem at hand as well as to prevent the same from reoccurring in the future.

6.1 Creating Public Awareness

At present, the public can be encouraged to limit the use of single-use masks and disposable gloves wherever possible. The use of the alternative washable and reusable cloth masks will instantly reduce the mass consumption of use-and-throw plastic based items, which will also help to reserve the limited surgical and N95 masks for people at the frontline. Washing of hands thoroughly with soap and water repeatedly is recommended than using disposable gloves. Policymakers may intervene to make sure that the mass supply of single-use medical or surgical masks should be limited and not readily available. Widespread awareness is also the need of the hour to alert the public of the dangers of indiscriminate disposal of PPE and masks. People need to adhere to the waste segregation guidelines responsibly, while the authorities need to impose strict penalties for negligence. Perhaps, this punitive measure would sensitize the public to cautiously dispose the right masks or PPE at the right coloured bin according to the recommended guidelines. Several guidelines are available to assist proper and effective disposal of waste, one of which is shown in Fig. 6. Citizens must do what they can in recycling containers and cans, cleaning them and keeping them aside for further processing, without mixing potential contagious disinfectants. Both the public and health centres can adopt the transformed disinfection strategy that is provided for radical reusage.

6.2 Futuristic Approaches

While the immediate approach of creating public awareness may plateau the heaping waste generation, however, a stringent strategy is required to combat the already generated waste and more importantly, to ensure that the rate inflow is reduced significantly in the coming times.

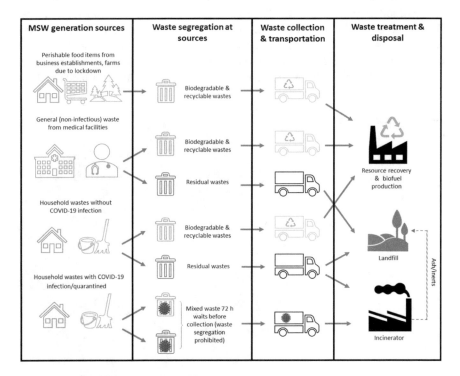

Fig. 6 Proposed MSW management [70]

6.2.1 Circular Economy Model

Lifecycle of plastic products must be taken care of by industries, ensuring their reusability and recyclability. Most systems adopt the linear economy model where the process flow, in most cases, leads to products ending in waste and the environment being ignored [71]. It is well known to all that our ecosystems are completely disrupted due to this 'use and throw' or 'take-make-use-dispose' concept. A revolutionized and more eco-friendly system known as circular economy model must, thus, be adopted to contain the influx of waste material (Fig. 7). This method enables businesses to control pollution, waste generation, reusability of products and regeneration of natural systems. It is believed that only a few retailers such as Adidas, Old Navy and Rent the Runway, as of October 2020, have adopted the model of circularity during the pandemic by creating reusable masks [72].

6.2.2 Use of Biodegradable Polymers

The choice of masks and PPE should be drawn towards biodegradable substitutes wherever possible. For instance, biodegradable nitrile gloves made from natural rubber proteins can be an alternative in non-hazardous contaminated environment such as food serving of COVID-19 patients and health workers. Polyvinyl alcohol (PVA) and polylactic acid (PLA) based gloves can also be used in dry conditions. A recent invention of reusable 3D printed PLA based N95 mask was developed, although its applicability in the long run is yet to be assessed [73]. PLA bags may also be used to contain dry waste. Interestingly, in 2014, Solubag, a biodegradable

Fig. 7 Circular business model

material for the creation of non-polluting bags, was invented that can dissolve readily in normal water within a few minutes. It is even claimed that the Solubag dissolved water is actually potable [74]. However, as expected, this product must be used only in dry conditions, which may restrict its suitability as PPE components. Nevertheless, research in this direction should be encouraged.

6.2.3 Policy Recommendations for Limiting Plastic Use

While the need of the hour is to prevent contagion by using more disposable inert plastic materials, it must be kept in mind that this 'relaxation' should not be the new norm. In the long run, policymakers must ensure that the unbridled use of plastic should be restricted and focus should be environment oriented. Proper reusing and recycling strategies for plastics must be thought of to ensure that the plastics do not end up polluting the environment. Another solution that has proven to work in countries like Ireland and China, is to impose high tax on single-use plastic bags and disposables purchases.

WHO, on its interim guidance on rational use of PPE, recommended minimal usage by restriction of unnecessary exposure, use of glass or window physical barriers, restriction of entry to contagious rooms of both healthcare workers and visitors, designating PPE only to a section of health workers who are at risks of exposure and so on [4]. Although the recommendations are listed in view of the current global short supply of PPE, however, logic dictates that the same rules are applicable even more so after the pandemic when the risk level of exposure to contagious infection is less. This will prevent misuse of PPE and subsequently, its disposal. Silva et al. proposed policy solutions to plastic and PPE waste management focusing on the possibility of decoupling plastic from fuel-based resources and optimising waste management [75]. Since plastic pollution is a global concern and is not limited to a certain territory, the authors believe that nations should cooperate in terms of knowledge, resources and funding. Since the problem is interdisciplinary, a tight collaboration between non-government organizations, academia, government bodies and stakeholders on addressing the need for new revolutionized approaches and policies must be enacted at the earliest.

6.2.4 Social Responsibility

A few organizations cannot fight the problem alone. However, the result of the small efforts of everyone is much more impactful. Especially in developing countries where people still irresponsibly throw garbage such as used masks and single-use plastics on the streets and open fields, awareness is a must. Consequences of this negligent act must be incited within the people through awareness programs, which is still lacking in many remote villages of many nations. Proper disposal and segregation of waste must be strictly adhered, steps that are still not followed sensibly even by the most educated section of the society in most countries like India. Unfortunately, most people, educated or otherwise, are unaware of the challenges of reckless waste management. They should be made conscious of the fact that the problem does not stop once the garbage is out of their homes. Heedfulness towards sanitary workers and waste-pickers must be instilled in the hearts of citizens through campaigns and social media to drive a humane attitude towards disposal of masks, PPE and other plastics. Every individual is also expected to follow the imposed guidelines of mask and PPE usage according to the area and country where one is residing, keeping in mind that a small act of responsibility can lead to a better tomorrow.

7 Summary

This chapter presents an overview of issues related to the recycling of facemasks and PPE. The outbreak of COVID-19 has caused unprecedented rise in the consumption of facemasks and PPE which are mostly manufactured by using non-biodegradable plastics. The existing BMW recycling plants are not enough to handle this surging amount of plastics. Though reuse of masks after autoclave and UV treatment is being explored, they are yet to be accepted as scalable technologies. Therefore, use of cloth masks that can be washed multiple times should be used where the risk of exposure is low. Creation of public awareness is of utmost importance in this regard as the segregation of waste at the origin is extremely crucial. Adoption of circular economy as a business model for facemasks and PPE can pave the way for complete reuse and recycling of these products.

References

1. Coronavirus disease 2019 (COVID-19) situation report—4. Accessed December 18, 2020. https://www.who.int/docs/default-source/coronaviruse/situation-reports/20200423-sitrep-94-covid-19.pdf
2. Coronavirus (COVID-19) events as they happen. Accessed December 18, 2020. https://www.who.int/emergencies/diseases/novel-coronavirus-2019/events-as-they-happen
3. Advice for the public. Accessed December 18, 2020. https://www.who.int/emergencies/diseases/novel-coronavirus-2019/advice-for-public/
4. World Health Organization (WHO) (2020) Rational Use of Personal Protective Equipment for Coronavirus Disease 2019 (COVID-19): Interim Guidance, 6 April 2020, vol 2019. https://apps.who.int/iris/bitstream/handle/10665/331695/WHO-2019-nCov-IPC_PPE_use-2020.3-eng.pdf
5. Prata JC, Silva ALP, Walker TR, Duarte AC, Rocha-Santos T (2020) COVID-19 pandemic repercussions on the use and management of plastics. Environ Sci Technol 54(13):7760–7765. https://doi.org/10.1021/acs.est.0c02178
6. COVID-19 has worsened the ocean plastic pollution problem—scientific american. Accessed December 18, 2020. https://www.scientificamerican.com/article/covid-19-has-worsened-the-ocean-plastic-pollution-problem/

7. Where did 5,500 tonnes of discarded face masks end up? Greenpeace International. Accessed December 18, 2020. https://www.greenpeace.org/international/story/44629/where-did-5500-tonnes-of-discarded-face-masks-end-up/

8. Opération Mer Propre. https://www.operation-mer-propre.com/?fbclid=IwAR1PxiR7kQBXs-HeNdOJIyW4iQUIKM39_vxv_MK82a91JO9pphl39g5w3cg

9. No shortage of masks at the beach—OCEANS ASIA. Accessed December 18, 2020. https://oceansasia.org/beach-mask-coronavirus/

10. Shortage of personal protective equipment endangering health workers worldwide. Accessed December 18, 2020. https://www.who.int/news/item/03-03-2020-shortage-of-personal-protective-equipment-endangering-health-workers-worldwide

11. COVID-19 Decontamination and Reuse of Filtering Facepiece Respirators|CDC. Accessed February 25, 2021. https://www.cdc.gov/coronavirus/2019-ncov/hcp/ppe-strategy/decontamination-reuse-respirators.html

12. Hart OE, Halden RU (2020) Computational analysis of SARS-CoV-2/COVID-19 surveillance by wastewater-based epidemiology locally and globally: Feasibility, economy, opportunities and challenges. Sci Total Environ 730:138875. https://doi.org/10.1016/j.scitotenv.2020.138875

13. NIOSH guide to the selection and use of particulate respirators certified under 42 CFR 84 1996. https://doi.org/10.26616/NIOSHPUB96101

14. Zhao M, Liao L, Xiao W et al (2020) Household materials selection for homemade cloth face coverings and their filtration efficiency enhancement with triboelectric charging. Nano Lett 20 (7):5544–5552. https://doi.org/10.1021/acs.nanolett.0c02211

15. Garrigou A, Laurent C, Berthet A et al (2020) Critical review of the role of PPE in the prevention of risks related to agricultural pesticide use. Saf Sci 123. https://doi.org/10.1016/j.ssci.2019.104527

16. MacIntyre CR, Seale H, Dung TC et al (2015) A cluster randomised trial of cloth masks compared with medical masks in healthcare workers. BMJ Open 5(4):e006577. https://doi.org/10.1136/bmjopen-2014-006577

17. Smith JD, MacDougall CC, Johnstone J, Copes RA, Schwartz B, Garber GE (2016) Effectiveness of N95 respirators versus surgical masks in protecting health care workers from acute respiratory infection: a systematic review and meta-analysis. CMAJ 188(8):567–574. https://doi.org/10.1503/cmaj.150835

18. Radonovich LJ, Simberkoff MS, Bessesen MT et al (2019) N95 respirators vs medical masks for preventing influenza among health care personnel: a randomized clinical trial. JAMA—J Am Med Assoc 322(9):824–833. https://doi.org/10.1001/jama.2019.11645

19. Klemeš JJ, Fan Y Van, Tan RR, Jiang P (2020) Minimising the present and future plastic waste, energy and environmental footprints related to COVID-19. Renew Sustain Energy Rev 127(April). https://doi.org/10.1016/j.rser.2020.109883

20. Managing infectious medical waste during the COVID-19 pandemic. Accessed December 18, 2020. https://www.adb.org/sites/default/files/publication/578771/managing-medical-waste-covid19.pdf

21. Delhi to Vijayawada, India has started dumping Covid-19 infected waste in public places. Accessed December 18, 2020. https://theprint.in/opinion/delhi-vijayawada-india-dumping-covid-19-infected-waste-public-places/496396/

22. Datta P, Mohi G, Chander J (2018) Biomedical waste management in India: critical appraisal. J Lab Physicians 10(01):006–014. https://doi.org/10.4103/jlp.jlp_89_17

23. Bio-medical waste management. Environ Inf Syst 2014:1–24

24. 10% of India's bio-medical waste in Delhi|Delhi News—Times of India. Accessed December 18, 2020. https://timesofindia.indiatimes.com/city/delhi/10-of-indias-bio-med-waste-in-delhi/articleshow/77116405.cms

25. Waste management in the context of the coronavirus crisis. Accessed December 18, 2020. https://ec.europa.eu/info/sites/info/files/waste_management_guidance_dg-env.pdf

26. Mobile medical waste incinerator sent to Wuhan to aid epidemic fight—Xinhua|English.news. cn. Accessed December 18, 2020. http://www.xinhuanet.com/english/2020-02/17/c_ 138792501.htm
27. Medical Waste Disposal—Definitive Guide 2020 [Infographic]. Accessed February 27, 2021. https://www.biomedicalwastesolutions.com/medical-waste-disposal/
28. Arvanitoyannis IS. Waste Management for Polymers in Food Packaging Industries. First Edit. Elsevier Inc. (2012). https://doi.org/10.1016/B978-1-4557-3112-1.00014-4
29. Waste incineration in Asia. PÖYRY. Published 2018. https://www.poyry.com/sites/default/ files/pov_asia01_2018_dec2018.final.pdf
30. Emmanuel J, Puccia CJ, Spurgin RA (2001) Non-incineration medical waste treatment technologies. http://goo.gl/QrK3hB
31. Feng S, Shen C, Xia N, Song W, Fan M, Cowling BJ (2020) Rational use of face masks in the COVID-19 pandemic. Lancet Respir Med 8(5):434–436. https://doi.org/10.1016/S2213-2600 (20)30134-X
32. Kashyap S, Ramaprasad A, Bidare SN (2020) Waste quarantine to reduce COVID-19 infection spread. Int J Health Plan Manage 35(5):1277–1278. https://doi.org/10.1002/hpm. 3026
33. No Govt Support, Maharashtra's Waste Pickers Struggle to Survive amid Fear of Covid Infection|NewsClick. Accessed December 18, 2020. https://www.newsclick.in/No-Govt-Support-Maharashtra-Waste-Pickers-Struggle-Survive-Fear-Covid-Infection
34. Beware the silent spreader—Covid waste—The Hindu BusinessLine. Accessed December 18, 2020. https://www.thehindubusinessline.com/news/beware-the-silent-spreader-covid-waste/ article32117688.ece
35. Gao QY, Chen YX, Fang JY (2020) 2019 Novel coronavirus infection and gastrointestinal tract. J Dig Dis 21(3):125–126. https://doi.org/10.1111/1751-2980.12851
36. Can Coronavirus spread through defective bathroom sewage pipes?|College of Public Health. Accessed December 18, 2020. https://cph.temple.edu/about/news-events/news/can-coronavirus-spread-through-defective-bathroom-sewage-pipes
37. Somani M, Srivastava AN, Gummadivalli SK, Sharma A (2020) Indirect implications of COVID-19 towards sustainable environment: an investigation in Indian context. Bioresour Technol Reports 11(June):100491. https://doi.org/10.1016/j.biteb.2020.100491
38. How long does PPE take to degrade naturally? The Eco Experts. Accessed December 18, 2020. https://www.theecoexperts.co.uk/blog/ppe-degradable
39. Ocean Plastics Pollution. Accessed December 18, 2020. https://www.biologicaldiversity.org/ campaigns/ocean_plastics/
40. Wilcox C, Van Sebille E, Hardesty BD, Estes JA (2015) Threat of plastic pollution to seabirds is global, pervasive, and increasing. Proc Natl Acad Sci USA 112(38):11899–11904. https:// doi.org/10.1073/pnas.1502108112
41. Pfaller JB, Goforth KM, Gil MA, Savoca MS, Lohmann KJ (2020) Odors from marine plastic debris elicit foraging behavior in sea turtles. Curr Biol 30(5):R213–R214. https://doi.org/10. 1016/j.cub.2020.01.071
42. 10-year wait for Delhi to clear its garbage piles?|Delhi News—Times of India. Accessed December 18, 2020. https://timesofindia.indiatimes.com/city/delhi/10-year-wait-for-capital-to-clear-its-garbage-piles/articleshow/78340326.cms
43. N95 Respirators, Surgical Masks, and Face Masks|FDA. Accessed December 18, 2020. https://www.fda.gov/medical-devices/personal-protective-equipment-infection-control/n95-respirators-surgical-masks-and-face-masks
44. Gandhi M, Beyrer C, Goosby E (2020) Masks do more than protect others during covid-19: Reducing the inoculum of sars-cov-2 to protect the wearer. J Gen Intern Med 35(10):3063–3066. https://doi.org/10.1007/s11606-020-06067-8
45. Science Review: September 26-October 2 2020|Prevent Epidemics. Accessed December 18, 2020. https://preventepidemics.org/covid19/science/weekly-science-review/september-26-october-2/

46. Chughtaita AA, Seale H, MacIntyre CR (2020) Effectiveness of cloth casks for protection against severe acute respiratory syndrome coronavirus 2. Emerg Infect Dis 26(10). https://doi.org/10.3201/EID2610.200948
47. When and how to use masks. Accessed December 18, 2020. https://www.who.int/emergencies/diseases/novel-coronavirus-2019/advice-for-public/when-and-how-to-use-masks
48. COVID-19: Should you switch your cloth mask for surgical or N95 masks?|The News Minute. Accessed December 18, 2020. https://www.thenewsminute.com/article/covid-19-should-you-switch-your-cloth-mask-surgical-or-n95-masks-135854
49. Cloth masks: if washed daily at high temperature, cloth masks can protect you from coronavirus: study—the economic times. Accessed December 18, 2020. https://economictimes.indiatimes.com/magazines/panache/if-washed-daily-at-high-temperature-cloth-masks-can-protect-you-from-coronavirus-study/articleshow/78641092.cms?from=mdr
50. Sterilizing surgical masks during a crisis. Accessed February 25, 2021. https://www.reviewofophthalmology.com/article/sterilizing-surgical-masks-during-a-crisis
51. van Doremalen N, Bushmaker T, Morris DH et al (2020) Aerosol and surface stability of SARS-CoV-2 as compared with SARS-CoV-1. N Engl J Med 382(16):1564–1567. https://doi.org/10.1056/nejmc2004973
52. Wuhui S, Bin P, Haidong K, Yanyi X, Zhigang Y (2020) Evaluation of heat inactivation of virus contamination on medical mask. J microbes Infect 15(1):31–35. https://doi.org/10.3969/J.ISSN.1673-6184.2020.01.006
53. Czubryt MP, Stecy T, Popke E et al (2020) N95 mask reuse in a major urban hospital: COVID-19 response process and procedure. J Hosp Infect 106(2):277–282. https://doi.org/10.1016/j.jhin.2020.07.035
54. Gilbert RM, Donzanti MJ, Minahan DJ et al (2020) Mask reuse in the covid-19 pandemic: Creating an inexpensive and scalable ultraviolet system for filtering facepiece respirator decontamination. Glob Heal Sci Pract. 8(3):582–595. https://doi.org/10.9745/GHSP-D-20-00218
55. Fisher EM, Shaffer RE (2011) A method to determine the available UV-C dose for the decontamination of filtering facepiece respirators. J Appl Microbiol 110(1):287–295. https://doi.org/10.1111/j.1365-2672.2010.04881.x
56. Zhang J-M, Zheng C-Y, Geng-Fu X, Yuan-Quan Z, Rong G (2004) Examination of the efficacy of ozone solution disinfectant in inactivating SARS Virus. Chin J Disinfect. Published online 2004. Accessed February 25, 2021. http://en.cnki.com.cn/Article_en/CJFDTotal-ZGXD2004 01010.htmDeğişikhastane,araştırmakuruluşuveüniversitearaştırmasonuçlarıdarefolarakverilmiş
57. Dennis R, Cashion A, Emanuel S, Hubbard D (2020) Ozone gas: scientific justification and practical guidelines for improvised disinfection using consumer-grade ozone generators and plastic storage boxes. J Sci Med 2(1). https://doi.org/10.37714/josam.v2i1.35
58. Grossman J, Pierce A, Mody J et al (2020) Institution of a novel process for N95 respirator disinfection with vaporized hydrogen peroxide in the setting of the COVID-19 pandemic at a large academic medical center. J Am Coll Surg 231(2):275–280. https://doi.org/10.1016/j.jamcollsurg.2020.04.029
59. Rubio-Romero JC, Pardo-Ferreira M del C, Torrecilla-García JA, Calero-Castro S (2020) Disposable masks: disinfection and sterilization for reuse, and non-certified manufacturing, in the face of shortages during the COVID-19 pandemic. Saf Sci 129(April):104830. https://doi.org/10.1016/j.ssci.2020.104830
60. Battegazzore D, Cravero F, Frache A (2020) Is it possible to mechanical recycle the materials of the disposable filtering masks? Polymers (Basel) 12(11):1–18. https://doi.org/10.3390/polym12112726
61. Zhong H, Zhu Z, Lin J et al (2020) Reusable and recyclable graphene masks with outstanding superhydrophobic and photothermal performances. ACS Nano 14(5):6213–6221. https://doi.org/10.1021/acsnano.0c02250
62. Summary for healthcare facilities: strategies for optimizing the supply of N95 respirators during shortages|CDC. Accessed December 18, 2020. https://www.cdc.gov/coronavirus/2019-ncov/hcp/checklist-n95-strategy.html

63. Chiang J, Hanna A, Lebowitz D, Ganti L (2020) Elastomeric respirators are safer and more sustainable alternatives to disposable N95 masks during the coronavirus outbreak. Int J Emerg Med 13(1):1–5. https://doi.org/10.1186/s12245-020-00296-8
64. "As if a storm hit": more than 40 Italian health workers have died since crisis began |World news|The Guardian. Accessed December 18, 2020. https://www.theguardian.com/world/2020/mar/26/as-if-a-storm-hit-33-italian-health-workers-have-died-since-crisis-began
65. NYC nurse who treated COVID-19 patients dies as one hospital reports 13 deaths in 24 hours —NBC New York. Accessed December 18, 2020. https://www.nbcnewyork.com/news/coronavirus/nyc-nurse-who-treated-covid-19-patient-dies-another-hospital-reports-13-deaths-in-one-day/2344831/
66. Chinese Center for Disease Control and Prevention Technical Guidance for Prevention and Control of COVID-19; 2020. Accessed December 18, 2020. http://www.chinacdc.cn/en/COVID19/202003/P020200323390496137554.pdf
67. Reusable Elastomeric Respirators in Health Care.; 2019. https://doi.org/10.17226/25275
68. Elastomeric respirators: strategies during conventional and surge demand situations|CDC. Accessed December 18, 2020. https://www.cdc.gov/coronavirus/2019-ncov/hcp/elastomeric-respirators-strategy/
69. Duling MG, Lawrence RB, Slaven JE, Coffey CC (2007) Simulated workplace protection factors for half-facepiece respiratory protective devices. J Occup Environ Hyg 4(6):420–431. https://doi.org/10.1080/15459620701346925
70. Kulkarni BN, Anantharama V (2020) Repercussions of COVID-19 pandemic on municipal solid waste management: challenges and opportunities. Sci Total Environ 743:140693. https://doi.org/10.1016/j.scitotenv.2020.140693
71. Vegter D, van Hillegersberg J, Olthaar M (2020) Supply chains in circular business models: processes and performance objectives. Resour Conserv Recycl 162(July):105046. https://doi.org/10.1016/j.resconrec.2020.105046
72. The growing impact of PPE & the waste issues COVID-19 has exposed. Accessed December 18, 2020. https://www.roadrunnerwm.com/blog/impact-of-ppe-waste
73. Jacob S, Joseph S, Menon VG (2020) Low cost preventative face shield and reusable N95 compatible mask—IEEE Future Directions. IEEE Future Directions. Published 2020. Accessed December 18, 2020. https://cmte.ieee.org/futuredirections/tech-policy-ethics/may-2020/low-cost-preventative-face-shield-and-reusable-n95-compatible-mask/
74. Solubag. Accessed December 18, 2020. https://solubag.cl/
75. Patrício Silva AL, Prata JC, Walker TR et al (2020) Rethinking and optimising plastic waste management under COVID-19 pandemic: Policy solutions based on redesign and reduction of single-use plastics and personal protective equipment. Sci Total Environ 742:140565. https://doi.org/10.1016/j.scitotenv.2020.140565

Assessment of Air Quality Impact Due to Covid-19: A Global Scenario

Snehal Lokhandwala, Dishant Khatri, and Pratibha Gautam

Abstract The global COVID-19 pandemic has put much of the world into lock-down which led to the unintended and positive changes in the environment surrounding us. It has benefited human race as it has led to improve air quality post-pandemic. Several studies state that the concentration of the various pollutants has decreased, where NO_2 air quality index value falls more precipitously (23–37%) relative to the pre-lockdown period, followed by PM_{10} (14–20%), SO_2 (2–20%), $PM_{2.5}$ (7–16%), and CO (7–11%), but the O_3 increases 10–27%. Due to this Covid-19 pandemic, D.C. has witnessed the cleanest spring since the last few decades, and L.A., one of the highly polluted city of the USA, has experienced improvement in the air quality. Air quality has always been an important issue where several big countries like China the COVID-19 outbreak has improved the air quality and lowered the pollution levels. According to several data analysis in the Chinese Ministry of Ecology and Environment, PM concentration levels decreased by more than 20% during January and April in more than 300 cities when compared to the previous month's pollution levels. There were many cities in which the air pollution level fell drastically during lockdown, 16% in Hong Kong, and 13% in Sydney and in 14% in Singapore. Few researchers studied that when restrictions were eased the PM 2.5 level increased for cities like Beijing, Los Angeles, Melbourne, Madrid, Cape Town and New York. There were cities like New York the pollution level took a dip during the peak lockdown by 59%. Also when talking about the Indian context, the air quality improved for several cities and states including Maharashtra being worst hit state. The air pollution level resulted in 'satisfactory' level (AQI) due to this lockdown. After this dip, there was again a peak at 33% increase in the air pollution when the city slowly started and reached to the pre-lockdown levels. This air quality has direct impact on the health of the people. Several air pollutants like NO_x have resulted on harmful effects such as increasing heart and lung disease. When examining the Indian population,

S. Lokhandwala (✉) · D. Khatri · P. Gautam
Department of Environmental Science and Technology, Shroff S R Rotary Institute
of Chemical Technology, UPL University of Sustainable Technology, Ankleshwar,
Gujarat, India
e-mail: snehal.lokhandwala@srict.in

S. S. Muthu (ed.), *COVID-19*, Environmental Footprints and Eco-design
of Products and Processes, https://doi.org/10.1007/978-981-16-3856-5_3

around 7% of total population is above the age of 65 and 27% in the age 0–14 who are more prone to the air pollution leading diseases like asthma, bronchitis, emphysema, and possibly cancer. This onset of pandemic has cleared the thought of people where immediate changes in air quality within dense population/industry can be improved based on pollution mitigation.

This chapter focuses on the worldwide air pollution studies and the findings can be implemented to maintain clean air. In addition, attempt has been made to assess the temporal behavior of daily Tropospheric Columnar NO_2 Flux and Ground-level NO_2 concentration for three days, one each in pre-lockdown, during lockdown and post-lockdown for three major cities of world, viz. New York, Mumbai and Wuhan. Also, NO_2 pollution assessment is done by mapping the remotely sensed columnar tropospheric NO_2 fluxes. Moreover, accelerated information approximately the hyperlink among air pollutants and COVID-19 can be useful worldwide with the aid of using informing public regarding fitness measures and sickness control techniques in scientific practice. The findings will be utilized by policymakers to set new benchmarks for air pollutants that could enhance the high-satisfactory life-styles for principal sectors of the World's populace and additionally to cut down the air pollutants in destiny with the aid of enforcing the strategic lockdowns on the pollutant hotspots with minimum financial drop.

Keywords Covid-19 · Air quality · Global scenario · Short-term and long-term exposures · Epidemiology · Temporal and spatial variability · Satellite studies · Pollutant concentration

1   Introduction

World's international populace has risen through 6.8 billion from last more than 20 decades. Researchers estimates that the populace could be 8.6 billion through mid-2030, more than 9.8 billion through mid-2050 and more than 11.2 billion through 2100. It is predicted to boom the Indian populace to attain 1.7 billion by 2050 [1]. The growth with inside the Urban Population and the subsequent 'Urbanization' comes particularly through changing the natural surfaces through the concrete systems and roads which alters the surface energy budget and outcomes within side the formation of increasingly more air pollutants emitting from expanded Industries in addition to increased automobiles on road. As a quit end result of urbanization and industrialization of human civilization, Air Pollution is considered as one of the major troubles in the twenty-first century posed to human beings. The nitrogen dioxide (NO_2) concentrations, resulting widely from the burning of fossil fuels [2], preceding to and following the quarantine, with a big bargain located in concentrations after the coronavirus outbreak [3]. The statistics had been collected with the resource of the usage of the Tropospheric Monitoring Instruments (TROPOMI) on-board ESA's Sentinel-5 satellite. NO_2 is a now no longer unusual place tracer of air pollution/business activity, associated with morbidity and mortality [2]. NASA scientists have commented that the reduction in NO_2 pollution modified into first appeared near Wuhan, but spread during the rest of the country, and eventually global [3]. In Central China, NO_2 emissions had been reduced with the resource of using as tons as 30% [3]. In Central China, NO_2 emissions were reduced by as much as 30% [3]. CO_2 emissions, another common tracer of air pollution, decreased by 25% in China and by 6% worldwide [4]. Air pollution is answerable for many deaths and extended incidences of respiration illness [5]. According to the World Health Organization, 4.6 million human beings die every year from ailments and illnesses at once related to bad air on first place [6]. Poor air is answerable for extra deaths each year than motor vehicle accidents [7]. The impact of air pollution is a global trouble and includes superior nations, which incorporates the European nations wherein 193,000 people died in 2012 from airborne particulate matter [8]. Air pollution associated deaths include but are not limited to asthma, bronchitis, emphysema, lung and coronary heart ailments, and respiration allergies [9]. Although the trendy implementation of emergency lockdown measures has contributed to a huge improvement in air over the area [10–13], the degrees of most air contaminants stay substantially higher than the values endorsed with the resource of the usage of the WHO in several nations [2, 55]. The rapid expansion of anthropogenic activities which incorporates transportation, business techniques and mining brought on a notable increase in plenty of risky pollutants that pose a chief risk to human health [15]. For instance, prolonged exposure to now no longer unusual place road way pollutants, which incorporates nitrogen oxides and ground-level ozone, can bring about oxidative stress and contamination within side the airways, inducing and drastically exacerbating health conditions principal to severe illness and pulmonary sickness leading to asthma, diabetes, cardiovascular disease [16]. These conditions have been tested to overlap with pathological abilities of COVID-19 lead illness, reinforcing the hypothesis of a dichotomy among air pollution and COVID-19 [17]. Nitrogen Dioxide (NO_2) is one of the highly reactive gases known as nitrogen oxides (NO_x). NO_2 widely gets within side the ambient surroundings via anthropogenic activities. Breathing air with an excess amount of NO_2 can bring about cardiovascular ailments. NO_2 and highly reactive NO_x interact with water, oxygen and highly reactive chemical materials within side the surroundings to form acid rain. Considering the big decrease in air pollution following the quarantine (China's CO_2 emissions decreased with the resource of the usage of a quarter), the COVID-19 pandemic could probable ironically have decreased the entire amount of deaths at some point of this period. Moreover, in addition to the reduced amount of deaths due to air pollution, the bargain in air pollution itself may additionally have remarkable benefits in reducing preventable non-communicable illnesses [18].

2 Air Pollution and Epidemiology

Air pollutants are not handiest a nearby hassle however is trans-boundary too as a result of the emission of positive pollution emitted from numerous specific locations both alone, or through chemical response causing terrible environmental and health influences that are both short and long term [7]. Along with urbanization there had been multiplied street automobiles that have been recognized as one of the critical members to air pollutants in metropolitan regions throughout the world, which could have an effect at the health of populace. Vehicular nitrogen oxide (NO_x) emissions immediately cause critical nitrogen dioxide (NO_2) pollutants troubles in the metropolitan regions, in which ambient NO_2 concentrations significantly exceed air quality standards. Furthermore, through complicated atmospheric techniques, automobile emissions have been the most important nearby pollutant to ambient high-quality particulate count (PM2.5) concentrations in large towns. Major reactive nitrogen compounds consisting of nitric oxide (NO), Nitrogen Dioxide (NO_2) and ammonia (NH_3) are critical precursor of concern in secondary particulate matter and ozone (O_3) [2]. Air pollutants can be classified commonly into 2 types: the ones having warming outcomes including especially Black carbon, and tropospheric ozone (O_3), and few having cooling outcomes including especially atmospheric aerosols composed of sulfur oxide (SO_x), nitrogen oxide (NO_x), particulate count (PM), natural carbon (OC), carbon monoxide (CO), non-methane unstable natural compounds (NMVOC), and ammonia (NH_3). The former are known as short lived weather pollution (SLCPs) due to their quick lifetime within side the surroundings which accounts from numerous days to a long time [4] while the Institute for Health Metrics and Evaluation divides air pollutants into 3 categories: ambient ozone pollutants, ambient particulate count pollutants, and household air pollutants from stable fuels. Sources of air pollution deterioration consist of motor automobile emissions, fuel generation, smelting and steel processing, several combustion, and others [18].

Air pollutants have now no longer only damaged human heath but surrounding also. There has usually been a want of having cleaner air which has additionally been known for a long time with movement having been taken at National and EU degree with additionally energetic participation at International conventions. There are numerous steps that had been taken which has decreased the air pollutants like emission from massive combustion plant and cellular sources, Fuel quality improved and environmental safety requirement fused into shipping and power [7]. WHO quoted that, 4.6 million people die yearly from illnesses and ailments immediately associated with bad air quality. It surpasses the deaths that is accounted via way of means of the street injuries every year. The effect of air pollutants is an international hassle and principal contributor is the evolved international locations, including the European international locations in which 193,000 human beings died in 2012 from airborne PM. Diesel engines emit surprisingly low concentrations of CO and CO_2, however in comparison with fuel engines of comparable size, diesel engines can generate range of debris with respect to distance traveled and are principal members to atmospheric PM concentration. In urban environment, 90% of traffic-generated PM is from diesel exhaust. In city areas, many city residents, consisting of people with multiplied susceptibility to the outcomes of air pollutants, have short-term exposure to diesel traffic during normal activities. Studies of people in exposure chambers have proven that managed exposure to diesel exhaust can initiate multiplied airway resistance and bronchial inflammatory changes. Such research are restricted via way of means of their artificial nature, however, and the small numbers of individuals have usually been healthy [19]. Air pollutants-related deaths consist of extensive horizon and now no longer restricted to annoyed asthma, bronchitis, emphysema, lung and coronary heart illnesses, and breathing allergies. The impact of air pollutants can be five million NCD (non-communicable deaths) deaths per year in which 90% of the world's populace is exposed to dangerous air. In relation to health, most concerning pollutants, are PM and Ground-level Ozone. Ground degree ozone is not shaped immediately however is shaped via way of means of the response of numerous VOCs and NO_x present within side the surroundings in addition to emitted from numerous sources [7]. While air pollutants are affected by individual choices to certain extent, shipping or different sorts of consumption-public policy has a critical element to play in shaping this principal determinant of health. NCDs related to air pollutants now no longer only effect the health and surroundings however additionally can have an effect on monetary boom in lots of ways. For example, when there are extra pollutants, there could be extra unhealthier populace and more individual gets ill and could be absent and much less active and productive. Recent research display the negative causal impact of short-run pollutants exposure on hours worked and productiveness at the same time as at work. An unhealthier populace that suffers from more air pollutants is likewise much less knowledgeable due to the fact kids with bad fitness have decrease school attendance rates and worse cognitive function [18]. Action that want to be taken to address these air pollutants calls for the involvement of now no longer simply the health sector however additionally the transport and power sector too. Local level adjustments are underway in a few towns across the world, including energy transition and concrete making plans that integrates extra green and much less polluting new transport solutions, so that it will be decisive for our health. If international locations and their public health groups are critical on having an effect on NCDs, these techniques want to be pushed via way of means of leadership from the health sector and additionally contribution from transport and energy sectors. Enhanced communication from the general public health network is fundamental to guide a shift in the direction of extra health-concentrated alternatives. WHO's BreatheLife marketing campaign and WHO's first Global Conference on Air Pollution and Health that was held in Geneva, Switzerland, Oct 30–Nov 1, 2018, have been critical to develop this agenda [20]. Similar form of conferences needs to propose concerning active participation and involvement of all of the stakeholders and coverage makers to make it more people's centric.

3   Covid-19 and Air Quality

Research across the globe and relevant experimental and epidemiological data suggest involvement of air pollution in outcomes related to Covid-19. Specific mechanisms have proved that air pollution plays a major role in Covid-19-related morbidity and mortality. Air pollution consist of complex mixture of gases and particulate matters that vary spatially and temporally [21].

3.1   Studies on Effect of Air Pollution on Prognosis of Covid-19

The impact of Covid-19 on air pollution varies across different areas which depend on time of virus introduction, the population density of area of impact, time of infection and control measures taken. Basic reproductive number (R_0) and effective reproductive number are used to estimate the influence of these factors on dynamics of the disease. These numbers depend on contact rates, transmissibility, duration of contagiousness and virus mutation time [22]. However, the impact of air pollution on spread of an infectious disease depends on several factors and very little is known till date for the plausible mechanism for the same. Some researchers believe that most of the indoor transmission is associated with droplets and patients infected with SARS-CoV-2 infection expel virus-laden droplets which can be controlled by physical distancing. This explanation does not hold true for short and long-range aerosols as they do not settle readily and thus the infected particles reaching lower airways contribute to increased frequency of more severe disease. Increased susceptibility to Covid-19 is associated with over expression of receptor for SARS-CoV-2 which is the result of chronic exposure to air pollution [23]. Several studies have analyzed the effect of air pollution on the prognosis of Covid-19. Data indicates that Covid-19 transmission is influenced by exposure to air pollution increasing vulnerability and harmful effects on prognosis of affected patients.

3.1.1   Short-Term Exposures and Temporal Variability

Models and Theory

There are numerous locations within side the city regions that often display networks which offer non-stop measurements of the attention of pollution at one or many locations, which may be used to examine the numerous topics. Such form of study and tracking information gives excessive stage of facts at the temporal variability in air pollutant attention and is extensively used to evaluate the effect of acute exposure on respiratory outcome inclusive of asthmatic symptoms [24], or persistent obstructive pulmonary ailment exacerbation [25]. These numerous research display the alternate within side the temporal modifications within side the publicity of the pollution. Such research usually use modeling method to account for recognized variability in exposure among numerous parameters that exist due to spatial variability or change in parameters, or the use of layout wherein parameters are limited to discrete region and activities [19]. The routinely available information in addition to the satellite-derived time series information and research related to it have performed a vital role in health impacts of air pollutants. These analyses make use of daily ground-level generated information to assess association with population-stage counts of particular health outcome inclusive of daily mortality [26], medical institution admission [27], or emergency room visit [28]. Various research usually average the ambient monitoring information for the study location and exposure on any given day is thought to be the identical for complete population. Although its miles a terrible assumption, the applicable parameters for interpretation are the quantity to which real private exposure are correlated to the location-averaged exposure over time. For Particulate Matter (PM), many research have advised fairly excessive correlations among city ambient concentrations and ground measured exposures [29]. For Ozone, these correlations are lower, while for each the instances stage of correlations varies throughout people and is possibly associated with the airflow residences of the indoor environment [30]. Also in numerous research, extra facts like private exposure to PM from non-ambient sources are likewise used for Particulate Matter (PM) [31, 32]. These analyses suggest that health implications are extra carefully associated with additives derived from the PM exposure from ambient source pollutants while in comparison with PM from indoor origin or measured overall private exposure (containing each additives). Although such research highlights the significance of indoor origin of exposure and indoor infiltration of outdoor pollutants, it is difficult to account for infiltration and individual instances region sample in studies of large populations. One new addition to study such issue is improvement and application of infiltration models such as CONTAM software [33], specifically for Particulate Matter (PM). With these models, building dynamic data and characteristics, derived from questionnaires or from numerous different ways [34], may be used to develop residence and season-specific factors to characterize infiltration. Such adjusting factors also can be carried out to research of exposure in conjunction with data on subject mobility.

Facts and Figures

- Among 4 ambient air pollutants (PM2.5, PM10, CO and NO_2), PM2.5 and NO_2 confirmed robust correlation with prevalence of Covid-19 in China. The Air great index (AQI) and prevalence of Covid-19 have been drastically correlated in Wuhan ($p < 0.05$) and Xiao gan ($p < 0.01$) in China [35].

- Setti et al. [36] gathered statistics from one hundred ten Italian towns and significant correlation among geographical distribution of daily exceeding PM10 values and preliminary spreading of Covid-19. The particulate matter serve as carrier for droplet nuclei enhancing the spreading of virus.
- A direct relationship was found among PM2.5 and PM10 to Covid-19 fatality in 3 towns of France [37].
- In China, the spreading of SARS-CoV-2 expanded through 5–7% for each 10 devices increase in AQI displaying a positive correlation among air pollutants indicator and confirmed Covid-19 cases [38].
- As studied through Frontera et al. [39], the maximum polluted areas in Italy have extra affected sufferers who required ICU admission. Moreover, those areas confirmed twofold excessive rate of mortality compared to other areas.

3.1.2 Long-Term Exposures and Spatial Variability

Models and Theory

The ambient tracking community information normally offer incredibly resolved and accurate information on temporal styles in air pollutants, it usually lacks precision concerning the spatial changes in the concentration in the urban area. This is an important shortcoming regard to chronic exposure, in which it's miles viable that specific houses/offices/place of work could have specific exposure to outside air pollutants that differs from others in same city location. Addition to this, use of chemical tracer techniques [40] enables to distinguish among sources contributing to the aggregate of pollution, temporal variability in concentrations is usually driven by the aid of using meteorological elements such that all pollutant aggregate, such as the ones from specific sources, rise and fall in concert. Some research that consists of distinctive spatial variability in awareness might also additionally permit source identification in certain areas (versus areas where sources do not contribute), permit for differentiation among sources contributing to exposure which may be used to evaluate the effect of long time exposure, that are considered as more impacting health. Mostly studies have been done recently about the long-term exposure focusing on the spatial comparison between cities with different ambient pollutant concentration with incorporating the individual level health information and other risk factors. Such approach, addition to creating our knowledge about knowing air pollutants health threat and improvement of regulations regarding it, mask any contribution of within-city differences in concentration, and with the aid of the effect of localized sources. In one study, the author made use of sources primarily based totally on an undertaking of topics to the display nearest to their house which improved the estimate of threat related to air pollutants, evaluating it with threat assigned to an area average [41]. While numerous research has determined bronchial allergies signs and symptoms, these findings are not consistent. Simple proximity measures would not calculate right sources as they lack statistics regarding traffic, vehicle, wind styles and numerous different topographical and meteorological information.

Facts and Figures

- 78% deaths in Covid-19 affected 5 areas of northern Italy and central Spain because of extended exposure to NO_2 found in higher concentrations combined with downward air pressure [42].
- A study in Italy said 9% boom in Covid-19 associated fatality in against one-unit boom in PM2.5 concentration [43].
- Data gathered from 355 municipalities in Netherlands confirmed nearly 100% boom in Covid-19 instances while pollutant concentrations elevated by 20% [44]
- A significant association among long-term exposure to PM2.5 and chance of deaths from Covid-19 become visible in national study in USA. The results prove that 8% boom in Covid-19 fatality rate with a boom of 1 μg/m3 in PM2.5 because of long-term exposure [45].
- Mele and Magazzino [46] said direct relationship among concentration of PM2.5 and Covid-19 mortality in a study covering 25 towns in India.

3.2 Effect of Covid-19 Containment on Air Pollution

The pandemic situation has reflected interdependence of human existence with environment and the ways to manage it. To combat this deadly virus, government across world have enforced variety of restrictions on mobility which has led to disruptions in economic activities. Several cities of globe including major cities of Europe, China, Brazil and India have noticed improvement in air quality as short-term effect of Covid-19 on natural environment [47, 48]. Estimates suggest that lockdowns and restrictions have prevented 0.65 million annual deaths for India [49]. Urban areas wherein more anthropogenic activities have shown a quantifiable reduction in major pollutants causing air pollution. A global study on emission of NO_2 and CO and Aerosol Optical Depth (AOD) showed a reduction of 0.0002 mol m^{-2} in NO_2, < 0.03 mol m^{-2} in CO and approximately 0.1–0.2 reduction in AOD at major hotspots between February–March as compared to 2019 [50]. Table 1 shows the effect of mitigation measures taken against Covid-19 and its impact on air pollutants.

Table 1 Effect of mitigation measures against Covid-19 and its impact on air quality in some areas of World

Area	Period of study (2020)	% reduction in			References
		PM2.5	NO$_2$	SO$_2$	
Northern Italy	Feb/March	–	30–40	–	Putaud et al. [51]
Europe	Feb/May	17	–	–	Giani et al. [52]
Ghaziabad, India	April/May	46.1	34.4	16.1	Lokhandwala and Gautam [47]
Scotland	March/April	6	40	–	Dobson and Semple [53]
Europe	March/April	–	5–55	–	Ordonez et al. [54]
20 cities of US	March/April	–	9.2–43.4	–	Goldberg et al. [55]
Yangtze river Delta region	Jan/March	22.9–54.0	–	–	Huang et al. [56]
Munnich, Germany	Entire 2020	–	24–36	–	Burns et al. [57]
Seoul and Daegu	Entire 2020	30.6–30.8	26.1–39.6	–	Seo et al. [58]
China	Jan/March	10.5	27	–	Silver et al. [59]

4 Satellite Studies: An Important Tool for Data Analysis for Impact Studies

Satellite information which is available in a big quantity which desires right information and efforts to carry new knowledge out of it. This satellite information in diverse codecs available may be a tool in order to permit efficient mapping and tracking of earth's resource, events, and ecosystem. This information/data may be used for diverse research which include administrative, clinical and industrial applications. This correct information permits the people to understand how we're affecting the nature and surroundings, which in turn permits information-based choices and actions. Access to satellite information permits to take well-timed movements and may understand what goes on small and big scales. Scientific Community points out that the information available is in so big amount that even 1% of the available information has not been in use.

The use of satellite information enables the government, industries and policy makers to take well-timed and higher choice and convey new improved services. The raw authentic information available in diverse codecs which may be interpreted using numerous remote sensing softwares. The different available parameters then may be blended and verified, with spatial, for further analysis. When activities, issues, changes, and developments may be detected, monitored and analyzed remotely with satellite information, the benefits for humans and environment may be tremendous.

There had been few new fields rising out and information available to discover the upcoming new fields and opportunities. The diverse fields wherein the satellite information are used:

- Crop Monitoring and future prediction of Agriculture
- Forestry Planning and Prevention of Illegal Logging
- Groundwater Monitoring
- Infrastructure Planning and Monitoring of constructing activity
- Coastal visitors tracking
- Studying the natural Catastrophe
- Climate change research
- Air pollutants research among few.

Satellite imagery databases provide low price satellite imaginary. Some of the information is likewise publicly available. Commercial satellite information companies sell satellite images, data and information, or they may be commissioned to provide continuous delivery of information as an example from a specific area. In general, the observations of remote sensing devices are quantitative. The raw facts are received at predefined coordinates and sensor specifications. The sensors record radiance, that's processed into raw photos for in addition evaluation and interpretation. In preprocessing image, co-registration and segment calibration are performed. For example, a SAR photo processing is composed of multiple stages, which include range compression after which azimuth compression. After that, the photos additionally are geo-referenced. Interpretation and evaluation of satellite data for satellite imagery are commonly carried out for the remote sensing software program or built-for-purpose algorithms. There are particular satellites used for particular data to be made available. Few satellites which are specifically for making available the pollutants near real-time associated information are:

- OMI (Ozone Monitoring Instrument)
- OPMS (Ozone Mapping and Profiler Suite)
- MODIS (Terra/Aqua)
- VIIRS
- AIRS (Atmospheric Infrared Sounder)
- TROPOMI (Tropospheric Monitoring Instrument)

Most of the information is to be had on the web platform freed from price which may be downloaded and used for numerous research and experiments. The most useful platform used for obtaining the information is NASAs Earth data. Each and every satellite has its own platform too from where the data may be acquired.

4.1 Case Study: Impact of Mitigation Measures During Covid-19 on Air Pollution of Wuhan, Mumbai and New York Using Satellite Data

4.1.1 Data Used

TROPOMI measures the Sunlight in 3 bands the ultraviolet and visible (270–500 nm), near-infrared (675–775 nm) and shortwave infrared (2305–2385 nm) spectral bands. The spectral resolution lies with inside the range of 0.25–0.55 nm. With a spatial decision as excessive as 7 km × 3.5 km and in a nadir pointing observation mode, it has the capacity to come across air pollutants over individual mega-towns. This is considerably smaller than its predecessor, OMI, which has a pixel length of around 24 km × 13 km, and clearly tons smaller than GOME-2 (80 km × 40 km) and SCIAMACHY (200 km × 30 km). The smaller pixel length approach air quality can likely be resolved on the size of mega-towns and distinguish specific industrial areas. The daily worldwide coverage is ensured with the aid of using a swath width of 2600 km. This additionally approach confirms additional measurements to be added to scientists, researchers. With TROPOMI, a brand new technology of demanding situations concerning large facts and the processing functionality is then open. It has a minimum lifetime of seven years. The TROPOMI tool is a multispectral sensor that gathers reflectance of wavelengths vital for measuring atmospheric concentrations of ozone, methane, formaldehyde, aerosol, carbon monoxide, nitrogen oxide, and sulfur dioxide, in addition to cloud traits at a spatial resolution of 0.01 arc degrees. The TROPOMI Level 2 processors will ingest Level 1B radiances with geolocations and irradiances. In addition, they may study auxiliary input data, each dynamic (e.g., meteorological fields from a numerical climate prediction model) and static (e.g., absorption cross-sectional reference spectra). Some Level 2 processors will ingest the Level 2 output from different algorithms. From those inputs, the processor will produce a TROPOMI Level 2 output record, as an example tropospheric NO_2 columns or O3 profiles. The processors may also produce a log file and an exit code in order that the processing machine can confirm that processing produced accurate results (Fig. 1).

The data used for this comparative analysis involves

- TROPOMI SENTINEL5P NO_x
- New dataset of with resolution of 5.5 km * 3.5 km

A database of daily columnar data of NO_2 from the Tropomi satellite available at a resolution of 5.5 km * 3.5 km was used for the study. The database was made by converting the NetCDF data files into.xlsx format using algorithms developed in Python programming language. The Database was then further trimmed and data points inside the study area were extracted. Mapping of the columnar NO_2 was done for various days to understand the spatial variability of the NO_2 over the city. The trimmed database of the columnar NO_2 was averaged which was then

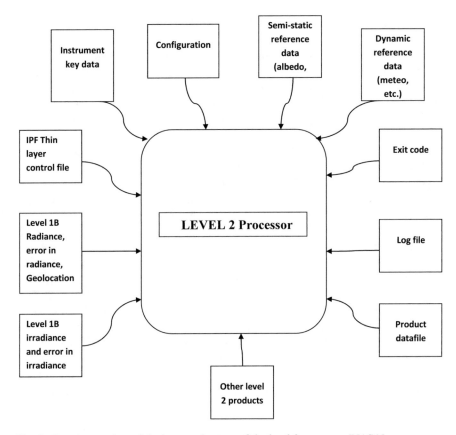

Fig. 1 Generic overview of the input and output of the level 2 processor [NASA]

considered as the representation for the city which is used to see the amount of change in the NO_x in the three different phase, viz. pre-lockdown, strict lockdown and post-lockdown phase.

Example of type of file created:

S5P_NRTI_L2__NO2____20181010T221303_20181010T221803_05144_01_010100_20
181010T225221.nc

S5P_OFFL_L2__NO2____20181010T225734_20181011T003903_05144_01_010100_20
181017T002032.nc.

4.1.2 Study Area and Methodology

There are many countries which are affected by the novel coronavirus out of which we have taken 3 sample cities, viz. Wuhan, New York, Mumbai to study the change in the NO_x level which can constitute a part of air pollution coming from various sources (Fig. 2). The pre-lockdown, Strict Lockdown and Post-lockdown phase has been studied.

Wuhan is the capital of Hubei province in China. It is the biggest town in Hubei and the maximum populous town in Central China with a populace of over 1.1 crore, it comes in the top 10 populous towns and ranks 9th among the national central city of China. Wuhan is in central-east Hubei, whose latitude and longitude ranges from 29° 58–31° 22 N and 113° 41–115° 05 E. Wuhan occupies a land place of 2,099,014.423 acremaximum of that has an alluvial plain and covered with natural surface like hills, water bodies and ponds. Water has been the main surface been covered in Wuhan's city territory which is considered best percentage among major towns in the China. Wuhan has 4 seasons with a humid subtropical climate with high amount of rainfall. Wuhan is understood for its humid summers, when dew points can frequently attain 26 °C (79 °F) or more.

New York City (NYC), also known as simply New York, is the maximum populous area in the US. With a 2019 populace of more than 83 lakh allotted over approximately 193,731 acre, located on the south most tip of New York is maximum densely populated city in USA, and the town is the center of the New York Metropolitan place, the biggest metropolitan place by urban available land. With nearly 2 crore humans in its metropolitan statistical area and about 2.3 crore in its blended statistical place, its miles one in all the worlds maximum populous megacity. The city of New York lies in between 40.730610 latitude, and the longitude 73.935242. The city's overall place is 299,520 acre; 193,280 acre of the town is land and 106,240 acre of that is water. Under the Koppen weather classification, the using the 0 °C (32 °F) isotherm, New York City has a weather which is humid subtropical weather, and is as a result the northernmost essential town at the North American continent with this categorization.

Mumbai also known as Bombay is the capital town of the Indian state Maharashtra. According to the UN, as of 2018, with the population of 2 crore Mumbai is the seventh most populous city worldwide and 2nd most populous city after the city of Delhi. Mumbai is on a narrow peninsula at the southwest of Salsette Island, which lies among the Arabian Sea to the west, Thane Creek to the east and Vasai Creek to the north. Mumbai's suburban district occupies maximum of the island. The overall place of Mumbai is 149,005 acre. The weather of the city is moist tropical and dry weather below Koppen weather classification… It varies among a dry length extending from October to May and a moist length peaking in June. The average annual temperature of 27 °C (81 °F), and the average annual precipitation is more than 2000 mm (85 in). In the Island City, the average most temperature is 31 °C (88 °F), even as the common minimum temperature is 24 °C (75 °F).

Methodology of the study involves following steps (Fig. 3).

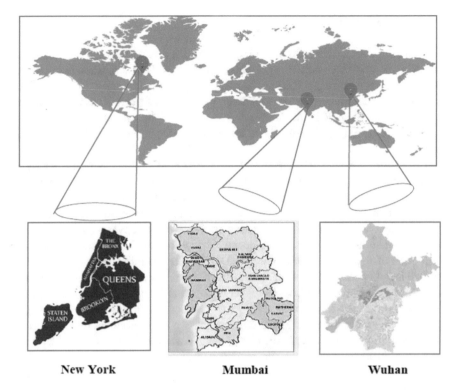

Fig. 2 Cities of world considered for study

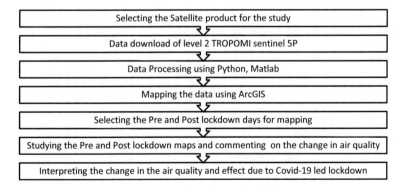

Fig. 3 Steps involved in interpreting data from level 2 TROPOMI sentinel 5P satellite

4.1.3   Results and Discussion

The data of Level 2 of the TROPOMI sentinel 5P for three cities Newyork, Mumbai and Wuhan was processed using Python, MATLAB. The data showing concentration of NO_2 was mapped using ArcGIS. Data of 17th January 2020 was considered as representative for pre-lockdown phase, 23rd March 2020 for lockdown phase and 5th July 2020 for post-lockdown phase. The nitrogen dioxide tropospheric column in ($*10^{-6}$ mol/m^2) for all three days for city of New York is shown in Fig. 4a–c.

Figure 4a for New York depicts that during the pre-lockdown period when there were active activities on a large scale then concentration of the Nitrogen dioxide in the area in and around New York was more which is shown in red color. The area showing more concentration are the center of the city. Figure 4b is the Lockdown period when due to more active cases in New York, government announced lockdown and then the phase wise opening which has led to decrease in the concentration of the Tropospheric Nitrogen Dioxide concentration. The concentration also changed because of the decreased industrial and other anthropogenic activities. Figure 4c depicts the Post-Lockdown period when there was partial opening of the activities which led to again increase in the Tropospheric Nitrogen Dioxide concentration which can be seen from areas which are again polluted due to deteriorating conditions in the city. In the city of New York from April to July 2020 there was four-phase reopening in the activities proposed by the government. The nitrogen dioxide tropospheric column in ($*10^{-6}$ mol/m^2) for all three days for city of Mumbai is shown in Fig. 5a–c.

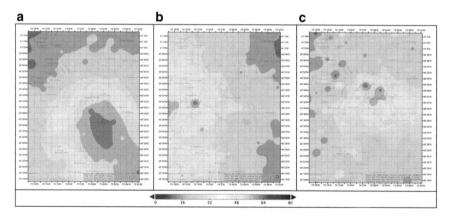

Fig. 4 **a** Satellite data-derived nitrogen dioxide tropospheric column map of New York showing the NO_x distribution on 17th January 2020 considering it as pre-lockdown day. **b** Satellite data-derived nitrogen dioxide tropospheric column map of New York showing the NO_x distribution on 23rd March 2020 considering it as lockdown day. **c** Satellite data-derived nitrogen dioxide tropospheric column map of New York showing the NO_x distribution on 5th July 2020 considering it as post-lockdown day

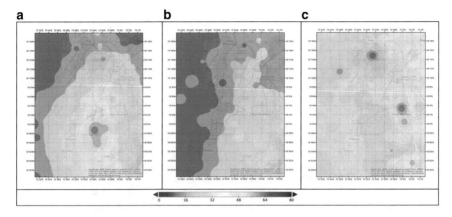

Fig. 5 **a** Satellite data-derived nitrogen dioxide tropospheric column map of Mumbai showing the NO_x distribution on 17th January 2020 considering it as pre-lockdown day. **b** Satellite data-derived nitrogen dioxide tropospheric column map of Mumbai showing the NO_x distribution on 23rd March 2020 considering it as lockdown day. **c** Satellite data-derived nitrogen dioxide tropospheric column map of Mumbai showing the NO_x distribution on 5th July 2020 considering it as post-lockdown day

Figure 5a for Mumbai depicts that during the pre-lockdown period when there were human activities on a large scale, the concentration of the Nitrogen dioxide in the area in and around the city center was more which is shown by red color. The area like Chinch Bandar which are main market hub with large number of vehicles lead to more concentration in the area. Lockdown was strictly imposed in Mumbai due to rapid increase in active cases in Mumbai and was declared as one of the hotspot in India for Covid-19. Connecting to this from Fig. 5b it is clearly seen that the Tropospheric Nitrogen Dioxide concentration decreases and pollution took a break. Figure 5c show re-emerging of red areas in the map which depicts the increase of Tropospheric Nitrogen dioxide again in post-lockdown phase.

The nitrogen dioxide tropospheric column in ($*10^{-6}$ mol/m^2) for all three days for city of Wuhan is shown in Fig. 6a–c.

Considering all the above maps 6(a, b, c) of the pre-lockdown, strict lockdown and post-lockdown phase, it is clearly seen that how because of the COVID-19 there are clear fluctuations in concentration of pollutants, in this case NO_x. This decrease is seen in most of the countries. The lockdown duration turned into complete of demanding situations and every people reacted to it differently. The duality of being quarantined: a few checked out it as a possibility at the same time as for a few others it turned into a burden. The most frequently mentioned benefit became the time that was freed through now no longer having to commute. This cover numerous aspects, including decreased traffic, extra free time, slowing down (now no longer having to hurry within side the mornings), or even decreased costs (because of now no longer the usage of the auto for work). Both, the perception of being quarantined and the coping techniques had been abundant;

a   b   c

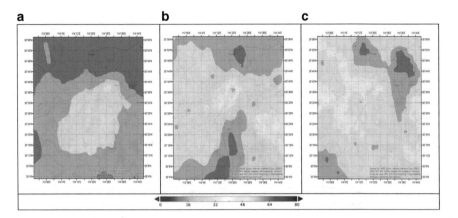

Fig. 6 **a** Satellite data-derived nitrogen dioxide tropospheric column map of Wuhan showing the NO_x distribution on 17th January 2020 considering it as pre-lockdown day. **b** Satellite data-derived nitrogen dioxide tropospheric column map of Wuhan showing the NO_x distribution on 23rd March 2020 considering it as lockdown day. **c** Satellite data-derived nitrogen dioxide tropospheric column map of Wuhan showing the NO_x distribution on 5th July 2020 considering it as post-lockdown day

however, it is clear that (growing the quantity of) telework—below regular circumstances—can be useful now no longer only to the surroundings but to the people as well.

5 Conclusion

The entire world is under strict regulations to combat the serious effect of air pollutants on human health but still there are numerous adverse effects that results from exposure of human beings to these agents. Epidemiological studies have clearly indicated a close association between the levels of air pollution with increased morbidity and mortality particularly due to number of respiratory diseases. Thus, it becomes very important to know how exposure to common air pollutants like NO_2, CO, PM2.5, PM10, etc. could exacerbate the susceptibility to and severity of virus infections. History says that pandemic situations have adverse effects on the human health and immediate antidotes are not available to nullify or cure the effect of the deadly virus. Thus, mode of social distancing, discipline in hygiene and forced lockdown are the preliminary ways to combat the contagiousness of Covid-19. The study in the chapter clearly indicates how the lockdown suppressed the concentration of NO_2 and ultimately improved the environmental conditions which were unfavorable for spread of Covid-19. Not only the immediate health impacts but the lockdown also changed lifestyle behaviors, mental health and economy. These indirect changes could surpass the observed burden of disease due to reduced air pollutant concentrations during lockdown periods. As such, there is no silver lining of effect of Covid-19 on the air pollution, but this study indicates that health benefits can be achieved due to lowered human and industrial activities ultimately lesser air emissions. Both mechanistic and epidemiologic studies including Satellite data are still required for proper understanding of how exposure to air pollutants could affect Covid-19 infections especially in populations already as hotspots of deadly SARS-CoV virus.

References

1. UN DESA (United Nations Department of Economic and Social Affairs) (2017) World population prospects: the 2017 revision. Retrieved from https://www.un.org/development/desa/publications/world-population-prospects-the-2017-revision.html#:~:text=The%20current%20world%20population%20of,Nations%20report%20being%20launched%20today. 21 June 2017
2. He L, Zhang S, Hu J, Li Z, Zheng X, Cao Y, Wu Y (2020) On-road emission measurements of reactive nitrogen compounds from heavy-duty diesel trucks in China. Environ Pollut 262:114280. https://doi.org/10.1016/j.envpol.2020.114280
3. NASA (2020) Retrieved from https://earthobservatory.nasa.gov/images/146362/airborne-nitrogen-dioxide-plummets-over-china
4. Hanaoka T, Masui T (2020) Exploring effective short-lived climate pollutant mitigation scenarios by considering synergies and trade-offs of combinations of air pollutant measures and low carbon measures towards the level of the 2 C target in Asia. Environ Pollut 261:113650. https://doi.org/10.1016/j.envpol.2019.113650
5. Brauer M (2001) Exposure of chronic obstructive pulmonary disease patients to particles: respiratory and cardiovascular health effects exposure of chronic obstructive pulmonary disease patients to particles: respiratory and cardiovascular health effects. J Expo Anal Environ Epidemiol 11(6):490–500. https://doi.org/10.1038/sj.jea.7500195
6. Cohen AJ, Brauer M, Burnett R, Anderson HR, Frostad J, Estep K, Forouzanfar MH (2017) Articles Estimates and 25-year trends of the global burden of disease attributable to ambient air pollution: an analysis of data from the global burden of diseases study 2015. Lancet 6736(17):1–12. https://doi.org/10.1016/S0140-6736(17)30505-6
7. Brussels (2005) Thematic strategy on air pollution, communication from the commission to the council and the European Parliament, Commission of the European Communities, 21st September, 2005
8. Ortiz C, Linares C, Carmona R, Díaz J (2017) Evaluation of short-term mortality attributable to particulate matter pollution in Spain. Environ Pollut 224:541–551. https://doi.org/10.1016/j.envpol.2017.02.037
9. Brauer M (2010) How much, how long, what, and where air pollution exposure assessment for epidemiologic studies of respiratory disease. Proc Am Thorac Soc 7:111–115. https://doi.org/10.1513/pats.200908-093RM
10. Bherwani H, Nair M, Musugu K, Gautam S, Gupta K, Kapley A, Kumar R (2020) Valuation of air pollution externalities: comparative assessment of economic damage and emission reduction under COVID-19 lockdown. Air Qual Atmosp Health 13:683–694. https://doi.org/10.1007/s11869-020-00845-3
11. Gautam S (2020) COVID-19: air pollution remains low as people stay at home. Air Qual Atmos Health 13:853–857. https://doi.org/10.1007/s11869-020-00842-6
12. Gautam S (2020) The influence of COVID-19 on air quality in India: a boon or inutile. Bull Environ Contam Toxicol 104:724–726. https://doi.org/10.1007/s00128-020-02877-y

13. Muhammad S, Long X, Salman M (2020) Science of the total environment COVID-19 pandemic and environmental pollution: a blessing in disguise? Sci Total Environ 728:138820. https://doi.org/10.1016/j.scitotenv.2020.138820
14. Wang P, Chen K, Zhu S, Wang P, Zhang H (2020) Resources, conservation & recycling severe air pollution events not avoided by reduced anthropogenic activities during COVID-19 outbreak. Resour Conserv Recycl 158:104814. https://doi.org/10.1016/j.resconrec.2020.104814
15. Bala R, Prasad R, Yadav VP, Sharma J (2018) A comparative study of land surface temperature with different indices on heterogeneous land cover using landsat 8 data. Int Arch Photogramm Remote Sens Spatial Inf Sci XLII-5:389–394. https://doi.org/10.5194/isprs-archives-XLII-5-389-2018
16. Guarnieri M, Balmes JR (2014) Asthma 1 outdoor air pollution and asthma. Lancet 383 (9928):1581–1592. https://doi.org/10.1016/S0140-6736(14)60617-6
17. Conticini E, Frediani B, Caro D (2020) Can atmospheric pollution be considered a co-factor in extremely high level of SARS-CoV-2 lethality in Northern Italy? Environ Pollut 261:114465. https://doi.org/10.1016/j.envpol.2020.114465
18. Chen S, Bloom DE (2019) The macroeconomic burden of noncommunicable diseases associated with air pollution in China. PLoS ONE 14(4):e0215663. https://doi.org/10.1371/journal.pone.0215663
19. McCreanor J, Cullinan P, Nieuwenhuijsen MJ, Stewart-Evans J, Malliarou E, Jarup L, Harrington R, Svartengren M, Han I-K, Ohman-Strickland P, Chung KF, Zhang J (2007) Respiratory effects of exposure to diesel traffic in persons with asthma respiratory effects of exposure to diesel traffic in persons with asthma. N Engl J Med 357:2348–2358. https://doi.org/10.1056/NEJMoa071535
20. Neira M, Pruss-Ustun A, Mudu P (2018) Comment reduce air pollution to beat NCDs: from recognition to action. Lancet 392(10154):1178–1179. https://doi.org/10.1016/S0140-6736(18)32391-2
21. Bourdrel T, Annesi-Maesano I, Alahmad B et al (2021) The impact of outdoor air pollution on COVID-19: a review of evidence from in vitro, animal, and human studies. Eur Respir Rev 30:200242. https://doi.org/10.1183/16000617.0242-2020
22. Heederik DJJ, Smit LAM, Vermeulen RCH (2020) Go slow to go fast: a plea for sustained scientific rigor in air pollution research during the COVID-19 pandemic. Eur Respir J 56:2001361
23. Paital B, Agarwal PK (2020) Air pollution by NO(2) and PM(2.5) explains COVID-19 infection severity by overexpression of angiotensin-converting enzyme 2 in respiratory cells:a review. Environ Chem Lett 19:25–42. https://doi.org/10.1007/s10311-020-01091-w
24. Liu L, Poon R, Chen L, Frescura A, Montuschi P, Ciabattoni G (2009) Research|children's health acute effects of air pollution on pulmonary function. Airway Inflamm Oxidative Stress Asthmatic Child 668(4):668–674. https://doi.org/10.1289/ehp11813
25. Liu Y, Yan S, Poh, K, Liu S, Lyioriobhe E, Sterling D (2016) Impact of air quality guidelines on COPD sufferers. Int J COPD 11:839–872. https://doi.org/10.2147/COPD.S49378
26. Zanobetti A, Schwartz J (2009) The effect of fine and coarse particulate air pollution on mortality: a national the effect of fine and coarse particulate air pollution on mortality: a national analysis. Environ Health Perspect 117(6):898–903. https://doi.org/10.1289/ehp.0800108
27. Tomic-Spiric V, Kovacevic G, Marinkovic J, Jankovic J, Cirkovic A, Milosevic Deric A, Relic N, Jankovic S (2020) Evaluation of the impact of black carbon on the Worsening of allergic respiratory diseases in the region of Western Serbia: a time-stratified case-crossover study. Medicina (Kaunas) 55(6):261. https://doi.org/10.3390/medicina55060261
28. Stieb DM, Szyszkowicz M, Rowe BH (2009) Air pollution and emergency department visits for cardiac and respiratory conditions: a multi-city time-series analysis. Environ Health 8:25. https://doi.org/10.1186/1476-069X-8-25

29. Sarnat JA, Koutrakis P, Suh HH (2011) Assessing the relationship between personal particulate and gaseous exposures of senior citizens living in baltimore. MD J Air & Waste Management Association 50(7):1184–1198. https://doi.org/10.1080/10473289.2000. 10464165

30. Kazakos V, Luo Z, Ewart I (2020) Quantifying the health Burden misclassification from the use of different PM 2.5 exposure tier models: a case study of London. Int J Environ Res Public Health 17(3):1099. https://doi.org/10.3390/ijerph17031099

31. Allen R, Wallace L, Larson T, Sheppard L, Sally Lui L-J (2007) Evaluation of the recursive model approach for estimating particulate matter infiltration efficiencies using continuous light scattering data. J Eposure Sci Environ Epidemiol 17:468–477. https://doi.org/10.1038/sj. jes.7500539

32. Strand M, Vedal S, Rodes C, Dutton SJ, Gelfand W, Rabinovitch N (2006) Estimating effects of ambient PM2.5 exposure on health using PM2.5 component measurements and regression calibration, 30–38. J Eposure Sci Environ Epidemiol 16:30–38. https://doi.org/10.1038/sj.jea. 7500434

33. Hystad PU, Setton EM, Allen RW, Keller PC, Brauer M (2009) Modeling residential fine particulate matter infiltration for exposure assessment. J Exposure Sci Environ Epidemiol 19:570–579. https://doi.org/10.1038/jes.2008.45

34. Gent JF, Koutrakis P, Belanger K, Triche E, Holford TR, Bracken MB, Leaderer BP (2009) Research|children's health symptoms and medication use in children with asthma and traffic-related sources of fine particle pollution 1168(7):1168–1174. https://doi.org/10.1289/ ehp.0800335

35. Li H, Xu X-L, Dai D-W, Huang Z-Y, Ma Z, Guan Y-J (2020) Air pollution and temperature are associated with increased COVID-19 incidence: a time series study. Int J Infect Dis 97:278–282. https://doi.org/10.1016/j.ijid.2020.05.076

36. Setti L, Passarini F, De Gennaro G, Barbieri P, Perrone MG, Piazzalunga A (2020) The potential role of particulate matter in the spreading of COVID-19 in Northern Italy: first evidence-based research hypotheses. Public Glob Health. https://doi.org/10.1101/2020.04.11. 20061713

37. Magazzino C, Mele M, Schneider N (2020) The relationship between air pollution and COVID-19-related deaths: an application to three French cities. Appl Energy 279:115835. https://doi.org/10.1016/j.apenergy.2020.115835

38. Zhang Z, Xue T, Jin X (2020) Effects of meteorological conditions and air pollution on COVID-19 transmission: evidence from 219 Chinese cities. Sci Total Environ 741:140244. https://doi.org/10.1016/j.scitotenv.2020.140244

39. Frontera A, Cianfanelli L, Vlachos K, Landoni G, Cremona G (2020) Severe air pollution links to higher mortality in COVID-19 patients: the "double-hit" hypothesis. J Infect 81:255–259. https://doi.org/10.1016/j.jinf.2020.05.031

40. Miller KA, Siscovick DS, Sheppard L, Shepherd K, Sullivan JH, Anderson GL, Kaufman JD (2007) Long-term exposure to air pollution and incidence of cardiovascular events in women. N Engl J Med 356(5):447–458. https://doi.org/10.1056/NEJMoa054409

41. Brunekreef B, Sunyer J (2003) Asthma, rhinitis and air pollution: is traffic to blame? Eur Respir J 21:913–915. https://doi.org/10.1183/09031936.03.00014903

42. Ogen Y (2020) Assessing nitrogen dioxide (NO_2) levels as a contributing factor to coronavirus (COVID-19) fatality. Sci Total Environ 726:138605. https://doi.org/10.1016/j. scitotenv.2020.138605

43. Coker ES, Cavalli L, Fabrizi E, Guastella G, Lippo E, Parisi ML (2020) The effects of air pollution on COVID-19 related mortality in Northern Italy. Environ Res Econ 76:611–634. https://doi.org/10.1007/s10640-020-00486-1

44. Andree BPJ (2020) Incidence of COVID-19 and connections with air pollution exposure: evidence from the Netherlands. Policy research working paper 9221, World Bank Group, fragility, violence and working theme, April 2020. https://doi.org/10.1101/2020.04.27. 20081562

45. Wu X, Nethery RC, Sabath BM, Braun D, Dominici F (2020) Exposure to air pollution and COVID-19 mortality in the United States: a nationwide cross-sectional study. Sci Adv 6:45. https://doi.org/10.1126/sciadv.abd4049
46. Mele M, Magazzino C (2020) Pollution, economic growth, and COVID-19 deaths in India: a machine learning evidence. Environ Sci Pollut Res 28:2669–6677. https://doi.org/10.1007/s11356-020-10689-0
47. Lokhandwala S, Gautam P (2020) Indirect impact of COVID-19 on environment: a brief study in Indian context. Environ Res 188:109807. https://doi.org/10.1016/j.envres.2020.109807
48. Zambrano-Monserrate MA, Ruano L, Sanchez A (2020) Indirect effects of COVID-19 on the environment Sci. Total Environ 728:138813. https://doi.org/10.1016/j.scitotenv.2020.138813
49. Sharma R, Singh SP, Bharti N (2019) Economic survey of Delhi 2018–19, Planning Department, Government of NCT of Delhi, February 2019
50. Lal P, Kumar A, Kumar S, Kumari S, Saikia P, Dayanandan A, Adhikari D, Khan M L, (2020). The dark cloud with a silver lining: Assessing the impact of the SARS COVID-19 pandemic on the global environment. Sci. Total Environ 732:139297. https://doi.org/10.1016/j.scitotenv.2020.139297
51. Putaud JP, Pozzoli L, Pisoni E, Santos SMD, Lagler F, Lanzani G, Santo UD, Colette A (2020) Impacts of the COVID-19 lockdown on air pollution at regional and urban background sites in northern Italy. Atmos Chem Phys Discuss 1–18. https://doi.org/10.5194/acp-2020-755
52. Giani P, Castruccio S, Anav A, Howard D, Hu W, Crippa P (2020) Short-term and long-term health impacts of air pollution reductions from COVID-19 lockdowns in China and Europe: a modelling study. Lancet Planet Health 4(10):E474–E482. https://doi.org/10.1016/s2542-5196(20)30224-2
53. Dobson R, Semple S (2020) Changes in outdoor air pollution due to COVID-19 lockdowns differ by pollutant: evidence from Scotland. Occup Environ Med 77:798–800. https://doi.org/10.1136/oemed-2020-106659
54. Ordonez C, Garrido-Perez JM, Garacia-Herrera R (2020) Early spring near-surface ozone in Europe during the COVID-19 shutdown: meteorological effects outweigh emission changes. Sci Total Environ 747:141322. https://doi.org/10.1016/j.scitotenv.2020.141322
55. Goldberg DL, Annenberg SC, Griffin D, Mclinden C, Lu Z, Streets D (2020) Disentangling the impact of the COVID-19 lockdowns on urban NO_2 from natural variability. Geophys Res Lett 47(17):e2020GL089269. https://doi.org/10.1029/2020GL089269
56. Huang L, Lui Z, Li H, Wang Y, Li Y, Zhu Y, Ooi MCG, An J, Snhang Y, Zhang D, Chan A, Li L (2020) The silver lining of COVID-19: estimation of short-term health impacts due to lockdown in the Yangtze River Delta region, China. Geohealth 4(9):e2020GH000272. https://doi.org/10.1029/2020GH000272
57. Jacob B, Sabine H, Christoph K, Michael L, Stephanie P (2021) Eva Rehfuess,COVID-19 mitigation measures and nitrogen dioxide—a quasi-experimental study of air quality in Munich, Germany. Atmos Environ 246:118089. https://doi.org/10.1016/j.atmosenv.2020.118089
58. Seo JH, Kim JS, Yang J, Yun H, Roh M, Kim JW, Yu S, Jeong NN, Jeon HW, Choi JS, Sohn JR (2020) Changes in air quality during the COVID-19 pandemic and associated health benefits in Korea. Appl Sci 10:8720. https://doi.org/10.3390/app10238720
59. Silver B, He X, Arnold SR, Spracklen DV (2020) The impact of COVID-19 control measures on air quality in China. Environ Res Lett 15:084021. https://doi.org/10.1088/1748-9326/aba3a2

Socio-economic Insinuations and Air Quality Status in India Due to COVID-19 Pandemic Lockdown

Meenu Gautam, Durgesh Singh Yadav, S. B. Agrawal, and Madhoolika Agrawal

Abstract In India, "Janta Curfew" with complete lockdown was imposed on 22 March 2020 to culminate the disease spread of COVID-19. Nationwide lockdown has created a socio-economic crisis, which has led to many human problems such as loneliness, mental depression, anxiety and domestic violence. The country witnessed the downfall in economic progress due to complete shutdown of industrial, commercial, transportation and other business ventures. On the contrary, lockdown facilitated our environment to restore its aesthetic value. The review beholds an objective to evaluate the impact of COVID-19 lockdown on socio-economic effects and air pollution level in India. The study includes comparative change in air quality during lockdown periods (22 March–14 April, 15 April–03 May, 4–17 May and 18–31 May 2020) against the same duration of 2019 and before lockdown periods in 2020. Air quality index was improved by >35% in 2020 compared to 2019 of same duration. Northern part of the country showed maximum reduction in air pollutants' level compared to southern part. Aerosol, $PM_{2.5}$, PM_{10}, NO_x, NO_2, NO, SO_2 and CO showed subsequent reductions in their levels during compared to before lockdown; however, O_3 level showed an increasing trend across the lockdown duration. Multivariate analysis showed a strong correlation of wind speed with total number of COVID-19 cases; however, ambient temperature exhibited a strong correlation with air pollutants. The chapter thus emphasizes upon the establishment and implementation of strict air control measures by government, policymakers and stakeholders with public participation for air quality improvement and long-term restoration of ecosystem in India.

Keywords Air quality index · Pollutants · COVID-19 · India · Lockdown · Meteorological parameters

M. Gautam · D. S. Yadav · S. B. Agrawal · M. Agrawal (✉)
Laboratory of Air Pollution and Global Climate Change, Department of Botany, Institute of Science, Banaras Hindu University, Varanasi 221005, India

© The Author(s), under exclusive license to Springer Nature Singapore Pte Ltd. 2021
S. S. Muthu (ed.), *COVID-19*, Environmental Footprints and Eco-design of Products and Processes, https://doi.org/10.1007/978-981-16-3856-5_4

1 Introduction

A novel coronavirus disease (COVID-19) was noticed in Wuhan city of Hubei province in China during December 2019 [29, 67]. The disease is caused by severe acute respiratory syndrome coronavirus-2 (SARS-CoV-2) [30, 65]. Spread of the disease occurred in the entire world and World Health Organization (WHO) declared the COVID-19 outbreak a global pandemic on 11 March 2020 [55]. There are a number of theories proposed regarding the origin of the novel coronavirus.

Accidental escape of the virus from the laboratory: The accidental escape of SARS-CoV-2 from the laboratory in Wuhan where researchers conduct experiments on bats' viruses for long time is often expected [2, 21, 42]. However, the origin and accidental escape of COVID-19 from China was denied based on the fact that genetic sequencing of new SARS-CoV-2 does not match any of the viruses sampled [2].

COVID was created as a biological weapon: [20] based on certain evidences stated that the virus is an offensive biological warfare weapon with DNA genetically engineered. But several studies debunked the theory based on the ground of genetic sequencing which reported that SARS-CoV-2 virus has entirely natural origin, i.e. as a zoonotic virus originating in bats [21, 53, 54].

Natural origin: Structural studies revealed that genome and spike protein present on the surface of SARS-CoV-2 cannot be genetically engineered but is possibly of natural origin [2, 22]. The virus showed >95% similarity with bat coronavirus and >70% homology with the SARS-CoV [45, 66]. Although many reports stated the origin of SARS-CoV-2 from bats, but pangolins and snakes are also suspected as intermediary animals through which it jumped into human acquiring genomic features under natural selection [2].

There are many more unresolved queries regarding novel coronavirus. However, concrete reason behind the origin of COVID-19 is yet to be found. As of now the world is more focussed on controlling the spread of the disease and introducing effective vaccines based on different technologies because the prevention time has already been passed.

1.1 Global and Indian Scenario of COVID-19 Cases

The contagious global pandemic, started in December 2019, has now been blown out to almost all countries. Total number of confirmed COVID-19 cases started increasing exponentially suggestive of the fact that human-to-human transmission was occurring [46]. The first fatality took place in China on 11 January 2020. For the month of January (Fig. 1), growth of confirmed cases was exponential whereas death cases took a steep rise from 26 to 31 January 2020 [56, 64]. In the month of February 2020, a logistic growth was found in total number of confirmed and death cases with subsequent increase in days (Fig. 1). March, 2020 was the most crucial period to take an action globally because the month showed an exponential rise in total number of confirmed and death cases due to COVID-19 (Fig. 1) [56, 64]. Exponential growth was prevailing as well as relentless phase for the large number of populations suffering from disease because in very short time number of infected person or fatality increased many folds [51]. Therefore, after declaration of the disease as pandemic by WHO, the entire world was imposed with strict norms and regulations: complete lockdown of transportation, industrial and commercial activities, avoidance of public gathering, religious activities, social distancing, frequent washing of hands, wearing masks and gloves, etc. Global shutdown shifted the exponential growth curve for confirmed and death cases (almost 1–9 folds increase in the number of cases with subsequent days) in March, 2020 to directly proportional growth curve in April and May, 2020 (1–4 and 1–2 folds increase in number of cases from the initial, respectively) [56, 64]. After May 2020, daily new cases increased till 7 January 2021 (maximum number 8444539) followed by a decline thereafter [58]. Similarly, total number of daily deaths after May, 2020 increased and was found maximum on 20 January 2021 (174,640) followed by a subsequent decline [58]. On March 2021, total number of confirmed COVID-19 cases has been reached to 125,496,328 globally, including 2,758,160 deaths and 101,351,863 recovered cases [56]. Top 25 countries such as the United States of America (USA) followed by the Brazil, India, Russia, United Kingdom (UK), France, Italy, Spain, Turkey, Germany, Colombia, Argentina, Mexico, Poland, Iran, Ukraine, South Africa, Czechia, Peru and Indonesia are the most vulnerable and lost more than 25,000 people till 25 March 2021 [56, 64].

In India, first COVID-19 patient was observed in Thrissur district of Kerala on 30 January 2020 [64]. Total number of confirmed cases in February reached to 3 with no causality (Fig. 1). On 3rd of March, 30 confirmed cases were reported mostly in Delhi, Jaipur and Agra due to foreign tourists and their contacts (www.corona.mygov.in). One case was reported in an Indian who travelled back from Vienna to Delhi and exposed number of school children in a party at the city hotel [45]. Alike global pattern, daily increase in total number of confirmed cases was exponential during March in India (approximately 51,555 and 200 folds increase in the number of cases by the end of first, second, third and fourth week of the month, respectively, from the initial) (Fig. 1) [64]. In April, sheerness in the exponential growth in number of confirmed cases with subsequent increase in days was slightly reduced (nearly 3, 6, 10 and 17 folds increase in the number of cases by the end of first, second, third and fourth week, respectively, from the initial). While in May, the growth curve for the cases turned out to be proportional (approximately 2, 2, 3 and 5 folds increase with the culmination of first, second, third and fourth week, respectively, from initial) with succeeding number of days [64]. Total number of daily cases increased proportionally after May, 2020 and maximum number (97,859) was found on September 16, 2020 followed by a decline till 8 February 2021 and again a marked increase was reported in total number of daily cases [64]. Furthermore, total number of daily deaths increased till 15 September 2020 followed by a decline up to 6 February 2021. Maximum number of death case (2006)

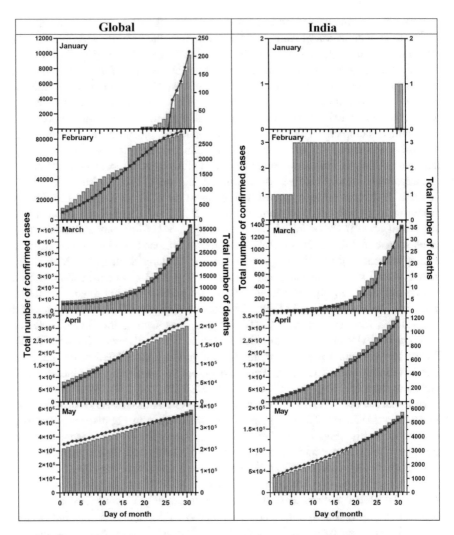

Fig. 1 Daily confirmed (bar) and death (line) COVID-19 cases in world and in India (1 Jan to 31 May 2020) based on World Health Organization [55–58] and WorldOMeter [63], respectively

so far due to COVID-19 was reported on 16 June 2020 [64]. On 31 March 2021, a total number of detected and death cases were 4,222,596 and 257, respectively [64].

1.2 Mitigation Strategies in India to Curb the Pandemic

Even though the mortality rate with SARS-CoV-2 is much lesser (4.33%) than the other pandemic diseases, i.e. SARS-CoV and Middle East respiratory syndrome

(MERS), but spread is relatively faster and 10–20 times more detrimental [26]. In view of this, Indian government started to form policies and regulations with strict laws to restrict COVID-19 pandemic at doorstep in India. Transportation and mass gathering were the prime reasons behind the rapid spread of the disease. Therefore, Indian government firstly came up with the issue of travel advisory guidelines on 11 January 2020 (Table 1). Subsequently, advisory guidelines were further upgraded and/or amended as per the enduring situation before nationwide lockdown (Table 1).

"Janta Curfew" imposed by the Government of India on 22 March 2020 appeared as a beginning of a long-term battle against the pandemic. Examining the public response of the "Janta Curfew", the country was then set to complete shutdown (Lockdown 1) from 25 March to 14 April 2020 with prior planning and extenuating measures (Table 1). Central armed police/paramilitary forces in assistance with state police forces were deployed within cities, at the airports, land ports, sea ports and borders to ensure the security of the country during the complete shutdown of the nation. Due to Lockdown 1 (L1), exponential growth in number of daily COVID-19 cases was though toned-down but the situation was yet out of control therefore the lockdown was further extended till 3 May 2020 (Lockdown 2 or L2). Second stage of lockdown came up with more strict rules and preventive measures to curb the pandemics (Table 1). During Lockdown 2, the worrying exponential growth curve of positive cases was changed to directly proportional growth curve but the situation was yet troublesome.

With the suggestions of state chief ministers and advisory committee, the nationwide shutdown was further extended from May 4 to 17 (Lockdown 3 or L3) with some additional rules and regulations (Table 1). Entire country was divided into red/hotspot (higher cases), orange (relatively less cases) and green (no positive cases) zones depending upon the number of COVID-19 cases. The health ministry has identified 319 districts as green, 284 as orange and 130 as red zones as of 4 May 2020 (www.corona.mygov.in). Few relaxations were given to orange and green zones, whereas red zone was debarred of such relaxations (Table 1). Hotspot areas were containment zones completely sealed and were under observation at regular time interval by health workers manually and through the app "Aarogya Setu and Ayush Kawach". Essential commodities to hotspot areas were facilitated by service providers.

Lockdown 4 or L4 (from 18 May to 31 May 2020) was further amended with change in regulations and additional relaxations to ease the public convenience (Table 1). Domestic flights were resumed with precautionary measures, groceries and other shops were opened, outpatient departmental (OPD) services in hospitals were resumed following preventive measures. However, COVID-19 hotspot areas were continued with strict regulatory measures. As of 31 May 2020, now there are 310 government laboratories and 111 private laboratories across the country are involved in SARS-CoV-2 testing (https://www.mygov.in/covid-19).

After Lockdown 4, unlock periods due to COVID-19 was divided into 4 time series from June to September, 2020 with different guidelines (Table 1). Besides, scientists were encouraged to develop the vaccines for COVID-19 with all the required support. Major biotech, i.e. the Serum Institute of India, Bharat Biotech, Premas Biotech and Zydus Cadila have been engaged actively in the vaccine

Table 1 Government strategies to alleviate COVID-19 pandemic in India

Date	Strategies
Before lockdown	
January 11	Avoidance of non-essential travels to China and Indians staying there are advised with following health measures: • Cover of mouth while coughing/sneezing • No travel plans when unwell and one should immediately seek prompt medical attention • Report to Indian Embassy in China in case of emergency
	On travelling back to India, if feel sick while flying: • One must inform the airlines' crew and follow the healthcare instructions • Report to the airport health and immigration offices at given numbers or e-mail addresses • Avoidance of close contact with family members and fellow travellers
	Travellers from China specifically Wuhan were required to follow health measures: • Personal hygiene, frequent hand washing, cover of mouth while coughing/sneezing • Symptomatic people were prohibited from social activities • Avoidance of travel to farms and live animal markets as well as eating raw/undercooked meats • Regular monitoring of their health
	Illness within a month after return from China: • Report to nearest health centres and helpline numbers and follow the healthcare guidelines • Self-isolation at home and with others, good personal hygiene, frequent hand washing and cover of mouth while coughing/sneezing
February 5	Indian travellers are advised to cease their travels from China but for those who were planning to return to India: • One with existing visas are ceased with further international travelling from China • People travelling to China were quarantined for at least 14 days on return • Intending visitors are obligatory to reapply for a fresh Indian visa • 24 × 7 helpline numbers and e-mail are provided by the Indian Embassy in China and Ministry of Health and Family Welfare (MoHFW), India for any assistance and help
February 26	• Indian citizens are advised to stop their travel to Singapore, Republic of Korea, Islamic Republic of Iran and Italy. Moreover, people who travelled from these countries to India were quarantined for 14 days
	• Travellers from above-mentioned countries were required to follow health measures as prescribed previously in travel advisory dated 11 February 2020
March 2–6	• Regular Visas granted to Indians/foreign national in China, South Korea, Japan, Iran, Italy, Hong Kong, Macau, Vietnam, Malaysia, Indonesia, Nepal, Thailand, Singapore and Taiwan, who have not yet entered India were suspended with immediate effect, while those who requires to travel to India due to compelling reasons have to reapply for fresh visa

(continued)

Table 1 (continued)

Date	Strategies
	• Diplomats, officials of the United Nations, OCI cardholders and other International bodies were allowed to enter India after their medical screening
	• Passengers travelling to India from any port were required to show self-declaration form, medical certificate of COVID-19 from the designated health authorities and travel history to medical and immigration officers at all port and must undergo medical screening at the port of entry
March 10	In addition to the travel advisory of 6 March 2020, the following were amended: • All incoming passengers to India should self-monitor their health and follow governments' instructions • Passengers from COVID-19 affected places were advised to undergo self-quarantine for 14 days from the date of their arrival but can work from home • Visas of foreign nationals (already in India) would remain valid and they may contact e-FRRO module for extension and conversion of their visa or grant of any consular services
	• International cruise ship from COVID-19 affected countries are either not permitted to enter any Indian port or facilitated with thermal screening in compulsory travel situation
	• No permission to symptomatic person to board and un-board the ship until tested negative for COVID-19. Person with COVID-19 positive are send to isolation facility
	• Regular sanitization of ship, provision of personal protection equipment and daily examination of passengers are accomplished
March 11	• International traffic through land borders are restricted at designated check posts with robust screening facilities
	• Indian nationals are strictly advised to restrict travelling to/from China, Iran, Italy, France, Germany Republic of Korea and Spain
	• All international passengers stepping into India were required to show self-declaration form (along with personal particulars to Health and Immigration officials. Also, they were supposed to undergo Robust Health Screening • For any queries related to health, people may contact Ministry of Health and Family Welfare at given 24 * 7 helpline number or e-mail
March 13	• Passengers' movement through immigration land check posts, i.e. Indo-Bangladesh, Indo-Nepal, Indo-Bhutan and Indo-Myanmar borders are suspended except for third country nationals' movement across Indo-Nepal and Indo-Bhutan borders after intensified health inspection and taking preventive measures as instructed by Ministry of External Affairs, Ministry of Home Affairs, Bureau of Immigration and MoHFW
March 14	• Prohibition of travel and trade from the European Union, UAE, Qatar, Oman, Kuwait, the UK and Turkey to India

(continued)

Table 1 (continued)

Date	Strategies
March 17	• Travel of passengers from Malaysia, Philippines and Afghanistan into India was prohibited with immediate effect
	• Ministry of Civil Aviation instructed the Airport Authorities to stagger the arrival of flights from COVID-19 affected areas to maintain the flow of passengers for medical screening
	• Arrived passengers were escorted by Airline Staff to the Association of Public Health Observatories • Health counters were installed to screen COVID-19 positive patients as per the existing norm
	• There are control room at the triage area of ports and five medical screening counters manned by medical officers and paramedical staff deputed by Delhi government • Passengers with no risks are advised for home quarantine after providing their passports, home quarantine advisory and collecting a declaration
	• The high-risk passengers before sending to quarantine centres are asked to fill up a declaration form opting for government/non-government facilities given by the state government
March 19	In continuation of the travel advisories issued on March 16–17, the following additional advisory was issued: • No scheduled international commercial passenger aircraft must take off from any foreign airport for any airport in India • Maximum travel time of 20 h is permissible for aircraft of commercial passenger to land in India
March 20	• Shipping services must be continued so that vital goods and essential commodities like fuel, medical supplies, food grains, etc. get delivered and economic activity of the nation do not get disrupted
	• Isolation/quarantine of symptomatic person in the ships' hospital
	• Vessels with infected person must be sanitized as per the protocols for dealing with COVID-19 pandemic
	• Vessels arriving in India from any port of COVID affected places have to undergo the necessary quarantine of 14 days
	• All ships' personnel who are likely to interact with the pilot must wear appropriate personal protective equipment (PPE)
	• Minimum person allowed to board the vessel
	• The master of the vessel must ensure the follow-up of the instructions issued by World Health Organization (WHO), International Maritime Organization (IMO), the MoHFW in India and other trade bodies
After nationwide lockdowns	
'Janta Curfew' + Lockdown 1 (March 22 to April 14)	• Complete shutdown of transportation and industrial activities except for essential goods i.e. medical and groceries
	• Educational institutes were closed
	• Government officials were allowed to work with 1 h in-between shifts following all safety measures

(continued)

Table 1 (continued)

Date	Strategies
	• Seamless supply of essential goods across the nation during the lockdown period
	• Relief funds for the poor and affected people and stimulus package to affected people during the lockdown
	• Train coaches and many educational premises were turned into isolation wards for patients of COVID-19
	• Cleaning and disinfecting public places at regular time interval
Lockdown 2 (April 15 to May 3)	• All domestic travel, passengers' train movement, metro rail services and other public conveyance were prohibited
	• Interdistrict and interstate mobility only allowed with medical reasons or emergencies. In case of two wheeler, driver and for four-wheeler, one passenger besides driver are allowed
	• Social/political/sports/entertainment/academic/cultural/religious functions/other public gatherings are paused
	• Demarcation of hotspot zones for COVID-19 following the guidelines of MoHFW
	• No unchecked inward/outward movement across the hotspot zones. Movement across the hotspot zones is allowed to maintain essential services such as medical emergencies, law and order related duties, groceries and sanitation facilities
	• Strict ban on the sale of gutka, liquor, tobacco, etc. Spitting should be strictly prohibited
	• All healthcare services (including AYUSH), medical research laboratories and sample collection centres to remain functional
	• Gathering of >5 people at public places is not allowed
	• In funerals, the congregation of >20 persons is not be permitted
	• Those industries or commercial activities involved in furnishing healthcare requirements/food supply/medical facilities, etc. are debarred from the rules
	• Financial sectors, fisheries, agricultural and horticultural activities, animal husbandry to remain fully functional
	• Operation of homes for disabled/children/senior citizens/mentally challenged/destitute/widows and women
	• Mahatma Gandhi National Rural Employment Guarantee Act (MNREGA) works must remain operative
	• Public utilities such as generation, transmission and distribution of electricity and fuels, postal services, telecommunications and Internet services must be operative
	• Movement of vehicles involved in loading/unloading of goods, cargo (inter- and intrastate) for fuels, food products, medical supplies and other essential goods
	• Government-approved Common Service Centres (CSCs) at Gram Panchayat level to remain functional

(continued)

Table 1 (continued)

Date	Strategies
	• Homestays, hotels, motels and lodges accommodated by tourists stranded in lockdown, medical/emergency staff, sea and air crew must remain open and functional
	• Construction of irrigation and renewable energy projects, roads, buildings and all kinds of industrial projects, including Micro, Small and Medium Enterprises (MSMEs) to be operational
	• Offices of the Government of India and its Autonomous/ Subordinate offices as well as State/Union Territory Governments and their associated bodies must remain open
	• Thermal screening at work place and proper sanitary habits, gap of 1 h. between the shifts at workplace, use of Arogya Setu app for working people and avoidance of large meetings
	• Person with comorbidities, above the age of 65 years and parents of children <5 years are encouraged to work from home
	• Proper disinfection of offices, factories and other establishments using user-friendly disinfectant mediums
	• Operation of 'Anganwadis'—distribution of healthcare items once in 15 days to beneficiaries
	• Online teaching/distance learning is encouraged
	• Medical insurance of workers
Lockdown 3 May 4 to 17	• Encourage to use staircase • Not beyond two or four persons in lift
	• Total ban on non-essential visitors
	• Segregation of places into red zones (high coronavirus cases), orange zones (relatively few cases) and green zones (no cases) • No free movements in red zone, use of only private/hired vehicles in orange zone for emergency purposes and use of public conveyance with 50% passenger boarding in green zone
Lockdown 4 May 18 to 31	Besides the advisories for Lockdown 3, following changes were made:
	• Shops opened from 10:00 AM to 5:00 PM following social distancing for buyers
	• Train/bus services for migrant workers to reach home safely • Stimulus packages for farmers, migrants and street vendors
	• Domestic flight services with 50% passengers' boarding, thermal screening at airports and following respiratory etiquettes
	• Permitted sale of gutka, liquor, tobacco, etc.
Lockdown 5 or Unlock 1 June 1 to 30	• Malls, hotels, restaurants and places of worship were reopen following guidelines laid by the government
	• Bharat Biotech announced it is the development of 'Covaxin' (India's first indigenous vaccine), in collaboration with ICMR and National Institute of Virology, Pune
Unlock 2 July 1 to 31	• Domestic flights and trains conveyance were resumed
	• Relaxation in night curfew

(continued)

Table 1 (continued)

Date	Strategies
	• Clinical trials of Covaxin began on 15 July
	• International flights were resumed after 'Bilateral Air Bubbles' with France and the USA
Unlock 3 August 1 to 31	• Gyms and yoga centres were allowed to reopen and night curfew was revoked
	• Indian drug firm Zydus Cadilla revealed the completion of Phase 1 clinical trial of its vaccine
	• The Indian Council of Medical Research reported more than one million diagnostic tests for COVID-19 had been collected
	• The Serum Institute of India also announced its clinical trials for 'Covishield'
Unlock 4 September 1 to 30	• Metro services resumed and schools were also partially reopened in several states
	• Dr. Reddy's Laboratories signed an agreement with the Russian Direct Investment Fund (RDIF) to conduct clinical trials of the Sputnik V vaccine in India
October 2020 to March 2021	• Decline in total number of COVID-19 cases
	• Union Health Secretary Rajesh Bhushan asked states and UTs to put in place a three-tier system that would oversee the rollout of COVID-19 vaccines
	• Pfizer and BioNtech announced that its vaccine candidate was more than 90% effective against COVID-19, according to results of Phase 3 trials
	• Reported daily recoveries were higher than daily new cases after 12-point plan (increasing number of ICU beds, oxygen cylinders, medical staffs and doubling the COVID-19 testing)
	• With Pfizer, Serum Institute of India and Bharat Biotech applying for emergency use authorization, India announced vaccine rollout plans and priority groups this month
	• A new and more infectious COVID-19 variant was detected in the UK and as a result India banned flights from the country to curb the spread, on 21 December
	• Night curfews in many states and UTs during New Year and Christmas celebrations
	• Schools and colleges were reopened following government guidelines
	• 51 lakh people will be vaccinated in the first phase in the national capital
	• Vaccination for health workers and people above 45 years of age through registration on COWIN and AAROGYA SETU applications
	• Increase in number of daily COVID-19 positive cases from February 2021

Source Compiled based on COVID-19 in India, Government of India (https://www.mygov.in/covid-19)

creation trails. The Serum Institute of India has a long history of producing vaccines against influenza, tetanus, measles, rabies and mumps. It has collaborated with Codagenix, a New York-based firm to develop a live-attenuated vaccine named ChAdOx1 nCoV-19 Corona Virus Vaccine (Recombinant) or COVISHIELD with 90% effectiveness against COVID-19 [41]. Other global companies actively involved in production of vaccines that are Pfizer-BioNTech, Moderna, Oxford AstraZeneca and Gamaleya (Sputnik V) with 95, 94.5, 82.4 and 91.6% efficacy, respectively, as per the report of March, 2021 [48]. The Covaxin by Bharat Biotech, India's first COVID-19 vaccine has demonstrated an interim efficacy of 81% in the phase 3 clinical trial and pose significant immunogenicity against variants [7]. These vaccines are meant to be given in two doses to the people at an interval of 4–6 weeks after registering on Cowin and Aarogya Setu applications. There were some common symptoms in 1 out of 10 people after the vaccination such as itching, pain, redness, swelling or bruising, tenderness and warmth, where the injection is given, feeling tired or unwell, headache, nausea and/or muscle ache [48]. 1 out of 100 people was observed with symptoms like feeling dizzy, poor appetite, abdominal pain, enlarged lymph nodes, excessive sweating and/or itchy skin. However, in severe allergic reactions, it has been advised to consult the doctor, healthcare provider or visit nearby hospital for proper medications. Even after the discovery of vaccines, many people with several misconceptions and medical complications are afraid of having vaccines which is creating difficulties in fighting against the disease. Other challenges are not following the government guidelines of social distancing and wearing masks which are responsible for increasing trend of active daily cases in India from February to March, 2021 [64].

The lockdowns due to COVID-19 pandemic enforce restrictions and self-quarantine measures, resulting in inordinate sufferings to mankind in terms of social well-being and economic development. However, this misfortune acted as "blessing in disguise" for our environment. Subsequent improvement in air and water quality was observed, noise pollution as well as waste generation was highly reduced [10] and vegetation cover across the nation was increased [36]. Amongst all, substantial decline in air pollution level during lockdown periods can provide an insight into the achievability of air quality improvement when there were reduced emissions from transportation, industries and other anthropogenic activities. In this chapter, we presented the socio-economic implications of COVID-19 pandemic and analysed the variations in meteorological as well as air quality parameters across different cities in India during lockdown periods.

2   Literature Survey and Assortment of Data

Secondary information relating to the study was facilitated by online sources. Daily situation reports on total number of COVID-19 cases worldwide and in India was incurred from the website of WHO (2020c) (https://covid19.who.int/) and WorldOMeter   (https://www.worldometers.info/coronavirus/),   respectively.

Information on economic growth rate in various sectors in India was facilitated through the website of Statista (2020) (https://www.statista.com/statistics) [46].

Meteorological parameters such as ambient temperature (AT), relative humidity (RH), solar radiation (SR) and wind speed (WS) from 21 different states including Andhra Pradesh, Assam, Bihar, Chandigarh, Delhi, Gujarat, Haryana, Jharkhand, Madhya Pradesh, Punjab, Rajasthan, Uttar Pradesh, West Bengal, Meghalaya and Mizoram (considered as northern part of India), and Karnataka, Kerala, Maharashtra, Odisha, Tamil Nadu and Telangana (considered as southern part of India). In each state, two cities were selected based on minimum and maximum population density from 1 January to 31 May 2019–2020 for the mean values of selected meteorological parameters and air pollutants [10]. Similarly, air quality index (AQI) and daily concentrations of eight air pollutants including particulate matter ($PM_{2.5}$ and PM_{10}), oxides of nitrogen (NO_x, NO and NO_2), sulphur dioxide (SO_2), carbon monoxide (CO) and ozone (O_3) during above-mentioned period were obtained from the two cities of each of the 21 states [10].

3 Statistical Analyses of Assorted Data

Linear regression analysis was performed between selected air pollutants ($PM_{2.5}$, PM_{10}, NO_x, NO, NO_2, SO_2, CO and O_3) against the period before (01 January to 21 March 2020) and during lockdowns ("Janta Curfew" + Lockdown 1 or L1: March 22 to April 14; Lockdown 2 or L2: April 15 to May 3; Lockdown 3 or L3: May 4 to 17; Lockdown 4 or L4: May 18 to 31, 2020). A linear regression analysis of air pollutants before and during the period of lockdowns was accomplished using GraphPad Prism (GraphPad Software Inc. version 8.0.1, San Diego, CA). For the multivariate analysis of meteorological parameters, air pollutants and total number of confirmed COVID-19 cases, principal component analysis (PCA) was performed. The whole data sets were analysed through PCA based on the correlation matrix with Varimax with Kaiser normalization. The multivariate test was performed by using SPSS software (IBM SPSS Statistics 20.0; IBM, Armonk, NY, USA).

4 Effects of Pandemic on Socio-economic Activities

4.1 Social Well-being

Symptoms of COVID-19 positive patients include headache, tiredness, aches/pains, nasal congestion, sore throat, dry cough, fever, conjunctivitis, loss of taste or smell, diarrhoea, discoloration of fingers or toes and rashes on skin [58]. Nearly one out of every five people who gets COVID-19 becomes seriously ill with symptoms of difficulty in breathing [58]. People with >60 years and those with underlying medical ailments like heart and lung problems, diabetes, high blood pressure and/or cancer are at a higher risk of developing serious illness [58]. Besides, increase in number of cases and fatalities due to the SARS-CoV-2; complete shutdown, social distancing, self-isolation were additively creating the health issues to human being. Self-isolation due to the disease causes chronic loneliness, boredom, anxiety and mass panic [5], wherein prolonged isolation leads to detrimental effects on physical and mental well-being (anxiety, depression, insomnia, chronic stress, adjustment disorder or even late-life dementia) [5]. Loneliness was amongst the most common adverse effects in the old-age group, leading to increased depression rate and sui-cidal tendencies. Irritability, frustration and obstruction in mental progression in children were situating their parents in trouble. Fear amongst health workers was leading them towards anxiety and frustration. The social restrictions cause other anomalies like boredom, anger, frustration and domestic interpersonal violence [49]. After the lockdown, even though the public places like shops, malls and temples are now open, but still there is fear of getting infected amongst people. Second wave of the COVID-19 pandemic has again knocked and the cases started rising in late February but the story changed in the second week of March and continued on a worse trajectory thereafter [23]. Moreover, Maharashtra, Karnataka, Punjab, Delhi, Gujarat and Madhya Pradesh are worse-affected states in the second wave of the COVID-19 pandemic due to which complete lockdown has again imposed in many cities within these states. People are advised to follow the social distancing and take all preventive measures to combat the disease. However, there are certain group of people who are bothering less and are less following the protective norms by the government.

The pandemic has disproportionately impacted the poor population in terms of livelihood and survival [4]. Many people have lost their jobs due to shutdown of industries/factories, private companies have ousted their workers to cope up with their financial crises, public violence for availability of limited resources, millions of migrant labourers, homeless individuals and daily wage workers are left with no option to stay at workplace and thus moving back to their native places. Shutdown of schools and colleges have badly affected the intellectual progression of students, generating the fear of their future progression and job accomplishments, etc. After the unlock procedure, companies and factories are now opened with the rules of limited number of employees on time slot basis. Schools from ninth class onwards and colleges for final year students are reopened.

4.2   Economic Development

India ranked 3rd in the world in terms of GDP, 16th in total exports and 11th in total imports during 2018 [37]. In 2019, India exported US$ 322.8 billion worth of goods around the world and gained the profit of 22.3% from 2018 to 2019. The country not only shares land and sea borders but also has trade relationship with Afghanistan, Bangladesh, Bhutan, Burma, China, Nepal, Pakistan, Indonesia, Maldives and Sri Lanka [37]. India's top listed trading partners are United States

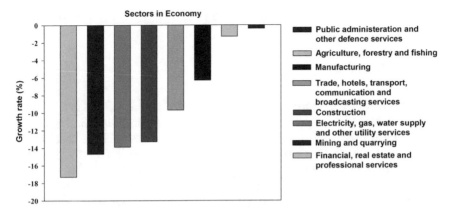

Fig. 2 Impact of nationwide lockdown due to COVID-19 pandemic on various sectors of economy in India compared to before lockdown based on data available at https://www.statista.com/statistics [46]

followed by United Arab Emirates, China, Hong Kong, Singapore, United Kingdom, the Netherlands, Germany, Bangladesh, Nepal, Belgium, Malaysia, Saudi Arabia, Vietnam and France [37]. Intercountries border shutdown and travel restrictions due to COVID-19 outbreak led to economic crisis and recession. The Indian economy is forecasted to shrink by 7% in this fiscal year as commodity export, demand, domestic supply, external financing, global trade and tourism have been disrupted due to the pandemic [61]. India's growth is estimated to have slowed to 4.2% in fiscal year 2019–2020 and from January to March 2021, the GDP of India has dipped to 3.2% [61]. To understand the turmoil effect of COVID-19 on the Indian economy, we summarized the negative effects on the growth rate of individual sectors including financial, real state and professional services ($\sim 17.3\%$), mining and quarrying ($\sim 14.7\%$), electricity, fuel, water and other utilities ($\sim 14\%$), construction ($\sim 13.3\%$), trade, transport, hotels and multimedia services ($\sim 9.7\%$), manufacturing ($\sim 6.3\%$), plantation and fisheries ($\sim 1.3\%$) and administration and defence services ($\sim 0.4\%$) (https://www.statista.com/statistics) [47] (Fig. 2). Preliminary assessment of COVID-19 impact on GDP by Maliszewska et al. [32] revealed the cumulative effect of −3.36, −0.84, −3.98, −8.23, −8.76 and −4.03% on agriculture, natural resources, manufacturing, domestic services and traded tourist services, respectively.

To subsidize the precarious situation of economy, Indian government, however, has taken appropriate steps to strengthen the economy, infrastructure, system, demography to meet the public necessities. There are policies, funds and planes to boost small and medium scale business/enterprises. Besides, government is encouraging to the use of domestic products that will help in maintaining the prosperity of Indian economy [14]. Fiscal incentive of Rs. 20 trillion has been allocated as a stimulus package for COVID-19 workers, migrants, medical facilities, employees and taxpayers, Micro, Small and Medium Enterprises (MSMEs), agriculture sector, tribal, farmers, Ministry of Finance, Government of India (2020).

5 Effects of Lockdown Due to COVID-19 on Meteorology and Air Quality

5.1 Meteorology

Diurnal variation in meteorological parameters including AT, RH, SR and WS from January to May 2020 is presented in Fig. 3. Before lockdown and during L1, L2, L3 and L4 periods, AT varied between 17.03–25.80, 24.95–28.85, 26.23–29.74, 27.55–29.52 and 28.17–31.78 °C, and RH varied between 55.03–76.71, 46.97–67.48, 50.09–65.92, 52.57–60.16 and 50.69–64.29%, respectively. Wind speed and SR during January 01 to March 21 (before lockdown) were in the range of 1.08–2.05 m $\sec^{-1}$ and 127.79–237.99 (W/m^2), respectively. While, WS and SR, respectively were in the range of 1.12–2.03 m $\sec^{-1}$ and 189.29–264.76 W/m^2 during L1, 0.97–1.75 m $\sec^{-1}$ and 186.6–234.79 W/m^2 during L2, 1.16–1.78 m $\sec^{-1}$ and 177.15–230.14 W/m^2 during L3 and 1.70–2.38 m $\sec^{-1}$ and 179.60 $-$229.62 W/m^2 during L4, respectively. Table 2 shows increase in RH, while AT was decreased before and during lockdowns compared to the same duration of previous year (2019). Maximum reduction in WS was found during L2 followed by L3 and L1 compared to the similar duration in 2019 (Table 2). However, SR showed variable trend across the lockdown duration.

5.2 Air Quality

Air pollution has been of mounting concern all over the world and specially in developing nations like India which ranked 5th in world's top 10 most polluted countries [60]. Indian cities have constantly been in the top 20 most polluted cities across the world based on air quality index as per WHO and CPCB [12, 33]. Air pollution in India is one of the prime causes of major health issues such as skin disease, respiratory and heart problems. About 1 million people die due to unhealthy air every year in India [43]. Anthropogenic emissions of air pollutants ($PM_{2.5}$, PM_{10}, NO_x, NO_2, NO, SO_2, CO and O_3) are increasingly responsible for deterioration of ambient air quality and changing climatic conditions [12, 33, 43]. COVID-19 lockdown led to substantial decrease in AQI, PM_{10}, $PM_{2.5}$, NO_x, SO_2 and CO concentrations in 2020 compared to 2019 followed by an increase in their levels after nationwide lockdown [12, 27, 31].

Air quality index is a parameter that reveals daily air quality; it signifies how polluted or clean the air is in that particular region and what are the associated health effects. AQI is basically calculated based on five major pollutants including particulate matter ($PM_{2.5}$ and PM_{10}), ground-level O_3, CO, SO_2 and NO_x [17]. AQI value is in the range of 0–500; higher the value, greater is the level of pollutants in air [17]. Figure 4 shows the AQI of 2019 and 2020 during the period of lockdown in the entire country. Lockdown duration in 2020 showed significant ($p < 0.001$)

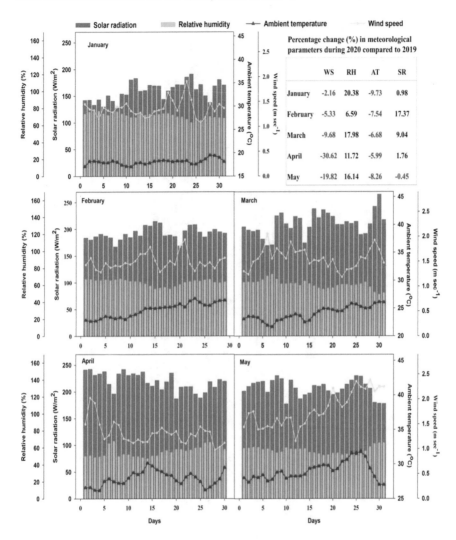

Fig. 3 Variations in selected meteorological parameters in India from 1 January to 31 May 2020 and their percentage change with respect to previous year based on data available at CPCB (2019–2020)

improvement in AQI compared to the same duration of previous year. Singh and Chauhan [44] also showed a pronounced decline in tropospheric NO$_2$ concentration, PM$_{2.5}$ and AQI over Chennai, Kolkata, Hyderabad, Mumbai and Delhi during the lockdown period in 2020 compared to 2019. Northern part of the nation displayed more reduction in AQI compared to southern part (Fig. 4). Sharma et al. [43] reported a decline of 29, 32, 44, 33 and 15% in east, west, north, south and central regions in India, respectively. The present study revealed that first, second,

Table 2 Percentage change (%) in meteorological parameters and the levels of air pollutants before and during lockdown periods in 2020 compared to the same duration of year 2019

Duration	Meteorological parameters				Air pollutants							
	AT	RH	WS	SR	$PM_{2.5}$	PM_{10}	NO	NO_2	NO_x	CO	O_3	SO
Before lockdown	−7.8	14.9	−6.0	7.6	−24.4	−23.1	8.9	−29.7	−5.2	−19.7	−15.1	−15.6
Lockdown 1	−7.0	8.2	−19.1	11.0	−38.5	−48.9	−45.4	−54.4	−58.4	−32.8	−3.8	−14.5
Lockdown 2	−6.6	18.6	−36.1	−3.7	−50.6	−48.7	−47.6	−50.1	−57.5	−39.6	−10.8	−10.4
Lockdown 3	−9.4	18.4	−33.6	−7.4	−46.6	−53.9	−31.1	−43.3	−47.1	−41.5	5.5	−7.6
Lockdown 4	−6.7	13.2	0.9	5.5	−36.6	−34.4	−46.3	−37.3	−48.5	−32.9	−7.7	−3.3

WS—wind speed; RH—relative humidity; AT—ambient temperature; SR—solar radiation; PM—particulate matter; NO—nitric oxide; NO_2—nitrogen dioxide; NO_x—oxides of nitrogen; CO—carbon monoxide; O_3—ozone and SO_2—sulphur dioxide (Based on data available at CPCB [10])

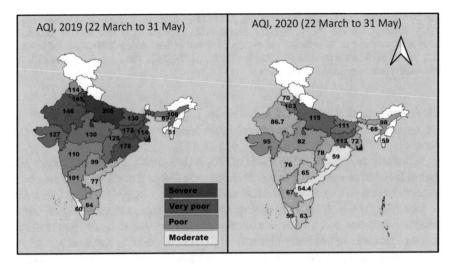

Fig. 4 National air quality index of different states of India during lockdown periods in 2020 compared to the same duration of 2019 based on data available at CPCB (2019–2020)

third and fourth lockdown duration resulted in 37.0, 39.6, 42.6 and 22.6% reduction in AQI compared to same duration of previous year (Fig. 4). Northern part of the country showed 36.4%, while southern part of the nation witnessed 33.9% decline in AQI (Fig. 4). Decline in AQI of the country is mainly ascribed to alteration in dominant pollutants such as PM_{10}, $PM_{2.5}$, NO_x and CO [28]. The dominant air pollutant in lockdown duration was O_3, while there were huge reductions in the levels of particulate matter and NO_x [25, 31, 43]. However, after lockdown period, deterioration in AQI of the country has been noticed and as per the present scenario many cities fall within moderate to poor categories [11]. Furthermore, Singarauli and Bhiwadi fall under very poor category based on AQI [11].

Particulate matter is the most overriding pollutant in major parts of the nation and is majorly contributed by energy generation, and vehicular, residential and industrial activities [13]. Significant reduction ($p < 0.001$) in particulate matter ($PM_{2.5}$ and PM_{10}) was observed before and during lockdown periods of 2020 against same duration of 2019 (Table 2 and Fig. 5). Furthermore, a drastic reduction in the level of PM_{10} was found during lockdown across the country which is mainly attributed to shutdown of their major contributory sources (Fig. 5). High reduction in the level of PM_{10} was found in northern compared to southern part of the nation. Furthermore, decrease in PM_{10} level was more during L2 followed by L3, L1 and L4, while in southern part, its reduction during lockdown periods was in order of L2 > L3 > L4 > L1 compared to before lockdown period. In northern part, substantial reduction in $PM_{2.5}$ concentration was observed compared to the southern part (Fig. 6). Furthermore, in northern region, higher reduction in $PM_{2.5}$ concentration was found during L2 followed by L3, L4 and L1 compared to before lockdown period. However, southern part of the nation showed higher reduction in

Fig. 5 Variations in particulate matter ($PM_{2.5}$ and PM_{10}), oxides of nitrogen (NO_x), nitric oxide ▶ (NO), nitrogen dioxide (NO_2), sulphur dioxide (SO_2), ozone (O_3) and carbon monoxide (CO) concentrations in India before and during lockdown periods during 2020 compared to same duration of 2019 based on data available at CPCB (2019–2020). The dashed lines indicate the 95%-confidence interval of the regression

$PM_{2.5}$ concentration during L4 followed by L2, L3 and L1 compared to before lockdown period. After lockdown period, the most prominent air pollutant across the nation is PM_{10} followed by $PM_{2.5}$ and O_3 [11].

Substantial reductions in the levels of NO, NO_2 and NO_x were found before and during lockdowns compared to the previous year (Table 2 and Fig. 5). Wherein, a huge reduction in the levels of NO_2 and NO_x were observed during lockdown periods (Fig. 5) which could be mainly attributed to complete shutdown of transportation facilities and industrial activities [43]. The decline in the pollutants in northern part was significantly higher ($p < 0.001$) compared to southern part of the country (Fig. 6). In former case, decline in the levels of NO, NO_2 and NO_x during the four stages of lockdowns was L1 > L2 > L3 > L4. Furthermore in southern part, the decline in NO level was maximum during L4 followed by L2, L3 and L1 (Fig. 6), while, L3 followed by L2, L4 and L1 showed high reduction in NO_2 level. The declining trend in NO_x level during lockdown periods was L4 (41.2%) > L2 (40.9%) > L3 (31.9%) > L1 (28.9%). After the unlock procedure across the country, a significant increase in the levels of NO_x specifically NO_2 has been found till now mainly accredited to reopening of industrial and transportation facilities [60].

Enormous reduction in the level of CO was observed across the period of nationwide lockdown in 2020 compared to the same duration in 2019 (Table 2 and Fig. 5). Northern part (36 to 41%) showed relatively higher reduction in CO level than in southern part (16–26%) of the country (Fig. 6). However, not much variation in its level was observed across the four stages of lockdowns in both the parts (Fig. 6), which could probably be attributed to the activities of biomass burning and coal based power plants [18]. For SO_2, reduction was very low in 2020 compared to the period in 2019 during the study period and also there was no prominent as well as definite trend for the above-mentioned period (Table 2 and Fig. 5). Northern part of the country was marked with less reduction in SO_2 level compared to the southern part (Fig. 6). In former case, the decline in SO_2 level was 4.7, 7.9, 6.7 and 15.3% during L1, L2, L3 and L4, respectively; whereas with increase in lockdown stages, percentage reduction in SO_2 level decreased from 42 to 16.3% compared to before lockdown period in southern part of the country. According to CPCB [11] report, a remarkable increase in the levels of CO and SO_2 was found till March where CO has been found to be fourth prominent air pollutant in the country mainly accredited to power plants and biomass burning.

Contrary to $PM_{2.5}$, PM_{10}, NO, NO_2, NO_x, CO and SO_2, O_3 showed significant ($p < 0.05$) increase in its level in both the years (2019 and 2020) of the same duration with high increment observed during lockdown periods (from 22 March 2020 onwards) (Table 2 and Fig. 5). During April to August, O_3 concentration is

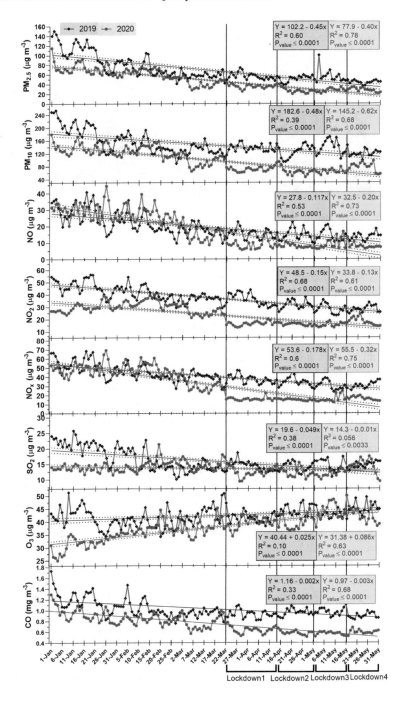

	Northern India				Southern India			
	L1	L2	L3	L4	L1	L2	L3	L4
PM2.5	-44.40	-59.55	-54.87	-52.05	-32.53	-48.16	-44.94	-51.26
PM10	-43.22	-46.55	-45.04	-26.19	-35.62	-50.29	-42.39	-41.70
NO	-79.10	-77.73	-77.13	-74.18	-10.99	-35.47	13.53	-43.36
NO2	-53.48	-52.79	-48.04	-40.34	-22.14	-34.10	-34.55	-22.63
NOx	-74.96	-73.92	-71.61	-67.02	-28.91	-40.86	-31.91	-41.24
CO	-36.12	-40.88	-40.23	-41.16	-15.66	-21.35	-26.78	-15.52
O3	18.95	18.57	44.46	30.01	6.76	-3.68	2.93	1.01
SO2	-4.68	-7.90	-6.69	15.31	-41.96	-40.24	-30.30	-16.26

Fig. 6 Percentage change in air pollutant ($PM_{2.5}$, PM_{10}, CO, SO_2, NO, NO_2, NO_x and O_3) levels in northern and southern part of India during lockdown periods compared to before lockdown based on data available at CPCB [10]

usually high, seemingly due to increase in solar radiation, industrial and transportation activities [19]. Unlike all air pollutants, increase in O_3 during study period may be ascribed to increase in hydrocarbons and decrease in NO_x concentration (a precursor of O_3 formation) [15]. In northern part of the country, maximum and minimum increase in O_3 level was found during L3 (44.5%) and L2 (18.6%), respectively, while in southern India, O_3 level was increased by 6.8, 3.0 and 1.2% during L1, L3 and L4, respectively (Fig. 6). However, during L2 duration, level of O_3 was decreased by 3.6%. This interpretation thus gives clear evidence that improvement in air quality can be accomplished with strict implementation of control measures. As per the present scenario, due to running industries, transportation facilities and increase in solar radiation, a significant increase in the level of O_3 (third most prominent air pollutant in the country) has been observed [11].

Changing meteorology plays significant role in pollutant formation [25]. Principal component analysis (PCA) was performed to assess the possible co-existence of any meteorological parameters (RH, AT, WS and SR) and air pollutants ($PM_{2.5}$, PM_{10}, NO, NO_2, NO_x, CO, SO_2 and O_3) (Fig. 7) which can be best correlated to COVID-19 dissemination in India between March 22 and May 31. Three principal components were obtained with 83.22% total variance. Component 1 (eigenvalue 5.9) contributed up to 45.65% of total variance, whereas component 2 (eigenvalue 2.8) and 3 (eigenvalue 2.1) elucidated 21.6 and 15.9%, respectively, of total variance (Fig. 7). Two groups were formed where first group showed close resemblance of ambient temperature with $PM_{2.5}$, PM_{10}, NO, NO_2, NO_x, CO, SO_2 and O_3. Jayamurugan et al. [24] also showed a strong positive correlation of particulate matter, NO_x, SO_2 and ambient temperature during summer season. In addition, Analitis et al. [1] reported synergistic effects of ambient temperature on O_3 and particulate matter. Second group elucidated a strong relationship between WS and total number of COVID-19 cases. Several studies showed a strong correlation amongst number of COVID-19 cases, AT and RH [3, 53, 54]. Auler et al. [3] through multivariate test showed a negative correlation between total

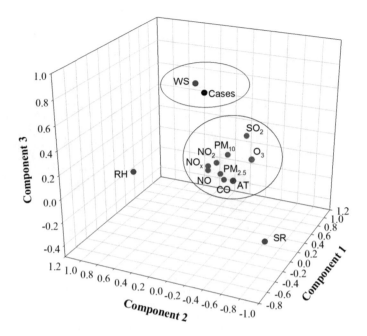

Fig. 7 Principal component analysis to assess the correlation amongst different meteorological parameters, air pollutants and total number of COVID-19 cases

number of COVID-19 cases in five Brazilian cities (São Paulo, Rio de Janeiro, Brasília, Manaus and Fortaleza) and temperature, while it exhibited positive correlation with rainfall, WS and RH. Moreover, temperature was seen to affect negatively the contamination rate and the number of cumulative cases, while relative humidity and wind speed had a positive influence during the present study (Fig. 7).

5.3 Case Studies

Ghaziabad followed by Delhi was the most polluted city in India based on environment performance index [62]. Nationwide lockdown due to COVID-19 witnessed a substantial minimization in the level of air pollutants in these two most polluted cities [10]. Ghaziabad situated in Uttar Pradesh of India has geographical area of about 777.9 km^2 and has a population density of 33,43,334 [6]. Sanjay Nagar of Ghaziabad, for an instance had AQI of 239 in 2019 and fell under the poor category [8]. Before and lockdown durations in 2020 cause significant reductions in AQI, $PM_{2.5}$, PM_{10}, NO_x, SO_2, CO and O_3 concentrations compared to previous year with few exceptions (Fig. 8) [10]. During first and second stages of lockdowns, AQI, $PM_{2.5}$, PM_{10} and NO_x showed maximum reduction compared to

Ghaziabad, Uttar Pradesh

Parameters	BL	During L1	During L2	During L3	During L4
AQI	35.3	-59.2	-61.4	-67.2	-22.9
		-65.4	-58.8	-56.4	-41.0
$PM_{2.5}$	25.0	-75.7	-55.0	-66.6	-26.2
		-79.7	-62.9	-64.5	-47.8
PM_{10}	6.5	-59.8	-62.5	-69.7	-43.4
		-51.1	-41.9	-41.5	-39.5
NOx	-8.6	-51.1	-57.0	-56.0	-32.7
		-50.1	-45.0	-31.0	-24.1
SOx	-52.5	-59.5	-68.9	-54.2	-11.9
		23.8	-13.5	31.3	78.5
CO	22.7	-60.0	-3.3	-2.0	14.0
		-55.4	-3.5	-36.3	-44.8
O_3	-55.6	-51.1	-55.1	-50.5	-45.7
		35.0	24.4	50.0	73.3

Sirifort, New Delhi

Parameters	BL	During L1	During L2	During L3	During L4
AQI	-9.1	-55.7	-45.1	-40.0	-16.3
		-57.2	-51.9	-27.6	-16.0
$PM_{2.5}$	-3.2	-65.7	-32.5	-50.9	-21.6
		-81.0	-57.8	-59.5	-55.6
PM_{10}	-21.6	-68.8	75.8	-62.1	-44.6
		-62.7	44.6	-43.1	-38.8
NOx	-7.4	-65.4	-48.9	-51.1	-26.6
		-68.9	-69.3	-45.3	-25.1
SOx	-40.2	-59.2	-7.8	44.6	77.8
		-24.3	-8.8	-0.3	1.8
CO	11.6	-17.2	35.4	-7.9	-50.7
		-56.7	-59.5	-64.9	-69.9
O_3	2.5	-59.1	-68.8	58.2	10.7
		-23.1	-32.6	150.3	190.2

Fig. 8 Percentage changes in AQI and air pollutant levels in Ghaziabad, Uttar Pradesh and Srifort, New Delhi (being highest polluted areas in India). BL, before lockdown; L1, lockdown 1; L2, lockdown 2; L3, lockdown 3; L4, lockdown 4; values in pink blocks represents percentage change within the duration of before and during lockdown periods of 2020 compared to 2019; values in green blocks represents percentage change in parameters during lockdown compared to before lockdown periods in 2020 based on data available at CPCB (2019–2020)

before lockdown duration (Fig. 8). SO_2 level was reduced during L2, whereas O_3 showed a subsequent increase in its level with increase in lockdown period after April 14. Concentrations of $PM_{2.5}$, PM_{10} and CO at Sanjay Nagar were above, while those of SO_2, NO_x, CO and O_3 were well within National Ambient Air Quality Standards (NAAQS) by CPCB [9] during the lockdown after April 14. Increase in $PM_{2.5}$ and PM_{10} concentrations could be ascribed to their high levels during early morning and late-night hours characteristically due to reduced ventilation and mixing height [10]. After the lockdown, $PM_{2.5}$, PM_{10}, NO_x, SO_2, CO and O_3 concentrations in Sanjay Nagar of Ghaziabad were in the range of 50–>400, 25–>400, 25–125, 0–25, 0–50, 0–125 from June, 2020 to March, 2021 [11].

The national capital territory, Delhi, is the second leading city in the world (The World's Cities 2018) with an area of about 1485 km^2 and 16.8 million residents (11,297 person/km^2) [6]. Shutdown due to pandemic led to reduction in AQI of the Delhi by 49% compared to before lockdown period [31]. Cumulatively, PM_{10} was decreased by 49% and NO_x by $\sim$ 10–70% in Delhi [40]. Sirifort, known to be the most polluted place in Delhi showed a significant reduction ($p < 0.001$) in AQI, $PM_{2.5}$, PM_{10}, NO_x, SO_2, CO and O_3 levels during the study period (before and during lockdowns) from 2019 to 2020 because the prime sources of these pollutants were restricted (Fig. 8). While, an increase in the level of PM_{10} during L2, SO_2 during L3 and L4, CO before lockdown and during L2, and O_3 before lockdown as well as during L3 and L4 were found in 2020 compared to the same duration of 2019 (Fig. 8). Maximum reductions in the AQI, $PM_{2.5}$, NO_x, SO_2 and O_3 levels were found during first followed by second lockdown periods, whereas CO showed a reverse trend of reduction as L4 > L3 > L2 > L1 compared to before lockdown period (Fig. 8). An increase in the level of PM_{10}, SO_2 and CO could be seemingly due to biomass burning and coal/gas based power plants [10]. Concentrations of $PM_{2.5}$ and PM_{10} (after April 14) as well as O_3 (after May, 3) did not comply with NAAQS by CPCB (2012). Sirifort which was under very poor category before lockdown based on AQI (>300), fell under moderate category (<100) due to COVID-19 lockdown [10]. From June, 2020 to March, 2021, the air quality index of Sirifort again deteriorated and fell under very poor category [11].

Aerosol is one of the important and diverse mixtures of air pollutants identified by international as well as national agencies [16]. It is tiny solid and liquid particles suspended in the air that reduce visibility and can cause heart and lung ailments [16]. Aerosol optical depth (AOD) is normally high in northern compared to southern part of the country [38, 50]. Urbanization, industrialization and other anthropogenic activities resulted in substantial increase in AOD in past few years [34]. COVID-19 lockdown has brought a remarkable change in the level of aerosol across the nation [52]. During the early spring period, urban areas of northern India produces lots of nitrates, sulphates, carbon-rich particles and soot through the vehicular discharge, industrial emissions and coal combustion [39]. Activities in rural areas such as cooking, heating stoves and farming fires add smoke rich in black and organic carbons [39]. Dust concentrations are generally low in March and early April, thereafter strong westerly winds blow sand from Thar Desert and Arabian Peninsula [34].

By all accounts, the 2020 lockdown reduced those human intervened emission sources and a remarkable decrease in AOD was found in the first week of the lockdown in combination with rainfall. Heavy rainfall around March 27 over vast areas of northern India helped in cleaning the air of aerosols (Fig. 9). AOD level in northern India at the beginning of April (by the end of L1) was significantly below the norm for this time of year (Fig. 9) and was lowest in 20 years of MODIS observations [34]. In southern India, satellite data shows that aerosol level has not yet decreased to the same extent, in fact, levels seem to be slightly higher than in the past four years whose probable reasons are still unclear but could be related to weather patterns, agricultural fires and wind pattern [34]. Unlock procedure and thereafter resulted in increase in AOD level across the nation specifically in northern part of the country compared to southern part from June, 2020 to February, 2021 [35].

6   Wake up Call for Government, Policymakers, Stakeholders and Common People

Following the statement that "adversity always introduces a man to himself", the pandemic situation has provided us an opportunity to understand that how capable and resilient we are in handling such extreme situations and also to comprehend the

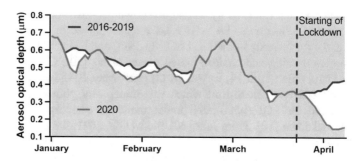

Fig. 9 Aerosol optical depth over northern India before and during lockdown (22 March to 7 April 2020) due to COVID-19 pandemic based on National Aeronautics and Space Administration [34]

importance of nature as well as natural resources. There could be long- and short-term effects of COVID-19 pandemic on our socio-economic activities in future but the pandemic has proven that the environmental deterioration caused by human intervened activities would consequently affect out survival in long term. The COVID-19 pandemic is an eye opener for everyone including government, policymakers, stakeholders and common people to plan strategies to come to normal life in post-pandemic era. Lockdown due to pandemic has brought miraculous change in environmental condition with an astounding improvement in air pollution level. But after the unlock procedure across the country, everything (industrial, transportation and other anthropogenic activities) came to the normal track and the pattern of environmental quality specifically air pollution level again worsened. Strict rules and regulations need to be imposed on anthropogenic activities which are the major causes of environmental degradation and disturbed ecology.

7 Conclusions

COVID-19 pandemic has adversarial impact on social well-being, human health and economy of India. However, 68 days of nationwide lockdown resulted in significant variation in meteorological parameters and improvement in air quality status. Wind speed and ambient temperature decreased while relative humidity increased before and after lockdown period in 2020 compared to the same duration in 2019. Substantial decline in the concentrations of particulate matters ($PM_{2.5}$ and PM_{10}), NO_x, NO_2, NO, SO_2 and CO was observed before and during four stages of lockdown periods in 2020 compared to similar duration in 2019. Maximum reduction in concentrations of PM_{10}, NO_x, NO_2 and CO was found, while O_3 showed increasing trend across the lockdown periods. Northern part showed greater reduction in air quality index ($\sim 36.4\%$) compared to southern part of the country

($\sim 33.9\%$). However, during the unlock procedure, air quality of the country reverted to its suspicious level. Multivariate analysis revealed a close resemblance of total number of COVID-19 cases with wind speed, thus suggesting that wind speed could be a possible factor in spread of the disease in Indian subcontinent. The study thus suggested that substantial environmental restoration can only be accomplished by strategic government policies, planned mitigative measure and more importantly public participation.

Acknowledgements Authors are thankful to the Head, Department of Botany, the Coordinator, Interdisciplinary School of Life Sciences and CAS in Botany, Institute of Science, BHU, Varanasi. Meenu Gautam is thankful to the Council of Scientific and Industrial Research (09/013(0857)/ 2018-EMR-I), New Delhi, for the financial aid in the form of Research Associateship. Durgesh Singh Yadav is thankful to University Grants Commission, New Delhi for the financial assistance in the form of SRF.

References

1. Analitis A, De'Donato F, Scortichini M, Lanki T, Basagana X, Ballester F, Astrom C, Paldy A, Pascal M, Gasparrini A and Michelozzi P (2018) Synergistic effects of ambient temperature and air pollution on health in Europe: results from the PHASE project. Int J Environ Res Public Health 15(9):1856. https://doi.org/10.3390/ijerph15091856
2. Andersen KG, Rambaut A, Lipkin WI, Holmes EC, Garry RF (2020) The proximal origin of SARS-CoV-2. Nat Med 26(4):450–452. https://www.nature.com/articles/s41591-020-0820-9.pdf
3. Auler AC, Cássaro FAM, da Silva VO, Pires LF (2020) Evidence that high temperatures and intermediate relative humidity might favor the spread of COVID-19 in tropical climate: a case study for the most affected Brazilian cities. Sci Total Environ 139090. https://doi.org/10.1016/j.scitotenv.2020.139090
4. Azeez EPA, Negi DP, Rani A, AP SK (2021) The impact of COVID-19 on migrant women workers in India. Eur Geogr Econ 62(1):93–112
5. Banerjee D, Rai M (2020) Social isolation in Covid-19: the impact of loneliness. Int J Soc Psychiatry. https://doi.org/10.1177/0020764020922269
6. Census of India (2011) http://census2011.co.in/
7. Chakraborty C, Agoramoorthy G (2020) India's cost-effective COVID-19 vaccine development initiatives. Vaccine 38(50):7883
8. CPCB (2019) https://app.cpcbccr.com/ccr/#/caaqm-dashboard-all/caaqm-landing
9. CPCB (Central Pollution Control Board) (2012) National ambient air quality status & trends in India, New Delhi
10. CPCB (2020) https://app.cpcbccr.com/ccr/#/caaqm-dashboard-all/caaqm-landing
11. CPCB (2021) https://app.cpcbccr.com/ccr/#/caaqm-dashboard-all/caaqm-landing
12. Das P, Mandal I, Debanshi S, Mahato S, Talukdar S, Giri B, Pal S (2021) Short term unwinding lockdown effects on air pollution. J Cleaner Prod 126514
13. Deng Q, Deng L, Miao Y, Guo X, Li Y (2019) Particle deposition in the human lung: Health implications of particulate matter from different sources. Environ Res 169:237–245. https://doi.org/10.1016/j.envres.2018.11.014
14. Dev SM, Sengupta R (2020) Covid-19: impact on the Indian economy. Indira Gandhi Institute of Development Research, Mumbai April. http://www.igidr.ac.in/pdf/publication/WP-2020-013.pdf

15. Dolker T, Agrawal M (2019) Negative impacts of elevated ozone on dominant species of semi-natural grassland vegetation in Indo-Gangetic plain. Ecotoxicol Environ Saf 182:109404. https://doi.org/10.1016/j.ecoenv.2019.109404
16. Dutheil F, Baker SJ, Navel V (2020) COVID-19 as a factor influencing air pollution? Environ Pollut. https://doi.org/10.1016/j.envpol.2020.114466
17. EPA (U.S. Environmental Protection Agency) (2016) Technical assistance document for the reporting of daily air quality—the AIR quality index (AQI). Publication No, Research Triangle Park, NC EPA-454/B-16–002
18. Gautam S (2020) The influence of COVID-19 on air quality in India: a boon or inutile. Bull Environ Contam Toxicol 1:724. https://doi.org/10.1007/s00128-020-02877-y
19. Gorai AK, Tchounwou PB, Mitra G (2017) Spatial variation of ground level ozone concentrations and its health impacts in an urban area in India. Aerosol and Air Quality Research 17(4):951. https://doi.org/10.4209/aaqr.2016.08.0374
20. Goyal VK, Sharma C (2020) The novel coronavirus 2019: a naturally occurring disaster or a biological weapon against humanity: a critical review of tracing the origin of novel coronavirus 2019. J Entomol Zool Stud 8(2):01–05. http://www.entomoljournal.com/archives/2020/vol8issue2S/PartA/SP-8-2-303-754.pdf
21. Graham RL, Baric RS (2020) SARS-CoV-2: combating coronavirus emergence. Immunity 52 (5):734–736. https://doi.org/10.1016/j.immuni.2020.04.016
22. Hassanin A, Grandcolas P, Veron G (2020) Covid-19: natural or anthropic origin? Mammalia, 1(ahead-of-print). https://doi.org/10.1515/mammalia-2020-0044
23. India Today (2021) https://www.indiatoday.in/coronavirus-outbreak/story/second-wave-coronavirus-10-points-1782146-2021-03-22. Accessed on 02/04/2021
24. Jayamurugan R, Kumaravel B, Palanivelraja S, Chockalingam MP (2013) Influence of temperature, relative humidity and seasonal variability on ambient air quality in a coastal urban area. Int J Atmos Sci. https://doi.org/10.1155/2013/264046
25. Kambalagere Y (2020) A study on air quality index (AQI) of Bengaluru, Karnataka during lockdown period to combat coronavirus disease (Covid-19): air quality turns 'Better'from 'hazardous'. Stud Indian Place Names 40(69):59–66. https://archives.tpnsindia.org/index.php/sipn/article/view/8200/7878
26. Krishnakumar B, Rana S (2020) COVID 19 in INDIA: strategies to combat from combination threat of life and livelihood. J Microbiol Immunol Infect 53(3):389–391. https://doi.org/10.1016/j.jmii.2020.03.024
27. Kumar S, Bhardwaj S, Singh A, Singh HK, Singh P, Sharma UK (2020) Environmental impact of corona virus (COVID-19) and nationwide lockdown in India: an alarm to future lockdown strategies. Preprints. https://doi.org/10.20944/preprints202005.0403.v1
28. Kumari P, Toshniwal D (2020) Impact of lockdown measures during COVID-19 on air quality—a case study of India. Int J Environ Health Res 1–8. https://doi.org/10.1080/09603123.2020.1778646
29. Lau H, Khosrawipour V, Kocbach P, Mikolajczyk A, Schubert J, Bania J, Khosrawipour T (2020) The positive impact of lockdown in Wuhan on containing the COVID-19 outbreak in China. J Travel Med 27(3):037. https://doi.org/10.1093/jtm/taaa037
30. Ma Y, Zhao Y, Liu J, He X, Wang B, Fu S, Yan J, Niu J, Luo B (2020) Effects of temperature variation and humidity on the death of COVID-19 in Wuhan, China. Sci Total Environ 138226:724. https://doi.org/10.1016/j.scitotenv.2020.138226
31. Mahato S, Pal S, Ghosh KG (2020) Effect of lockdown amid COVID-19 pandemic on air quality of the megacity Delhi, India. Sci Total Environ 730:139086. https://doi.org/10.1016/j.scitotenv.2020.139086
32. Maliszewska M, Mattoo A, Van Der Mensbrugghe D (2020) The potential impact of COVID-19 on GDP and trade: a preliminary assessment. https://elibrary.worldbank.org/doi/abs/10.1596/1813-9450-9211
33. Mukherjee A, Agrawal M (2018) Air pollutant levels are 12 times higher than guidelines in Varanasi, India. Sources and transfer. Environ Chem Lett 16(3):1009–1016. https://doi.org/10.1007/s10311-018-0706-y

34. NASA (National Aeronautics and Space Administration) (2020) https://earthobservatory.nasa.gov/images

35. NASA (National Aeronautics and Space Administration) (2021) https://earthobservatory.nasa.gov/images

36. NOAA (National Oceanic and Atmospheric Administration) (2020) Land surface temperature and vegetation. https://earthobservatory.nasa.gov/global-maps/MOD_LSTD_M/MOD_NDVI_M

37. OEC (The Observatory of Economic Complexity) (2020) https://oec.world/en/profile/country/ind/

38. Rajeev K, Ramanathan V, Meywerk J (2000) Regional aerosol distribution and its long-range transport over the Indian Ocean. J Geophys Res Atmos 105(D2):2029–2043. https://doi.org/10.1029/1999JD90041

39. Ramachandran S, Rupakheti M, Lawrence MG (2020) Black carbon dominates the aerosol absorption over the Indo-Gangetic Plain and the Himalayan foothills. Environ Int 142:105814. https://doi.org/10.1016/j.envint.2020.10581

40. Saxena A, Raj S (2021) Impact of lockdown during COVID-19 pandemic on the air quality of North Indian cities. Urban Clim 35:100754. https://doi.org/10.1016/j.uclim.2020.100754

41. Serum Institute of India Pvt. Limited (2021) https://www.seruminstitute.com/pdf/covishield_fact_sheet.pdf. Accessed on 04/04/2021

42. Shanker A (2020) The possible origins of the novel coronavirus SARS-CoV-2. https://doi.org/10.31219/osf.io/a83r4

43. Sharma S, Zhang M, Gao J, Zhang H, Kota SH (2020) Effect of restricted emissions during COVID-19 on air quality in India. Sci Total Environ 728:138878. https://doi.org/10.1016/j.scitotenv.2020.138878

44. Singh RP, Chauhan A (2020) Impact of lockdown on air quality in India during COVID-19 pandemic. Air Qual Atmos Health 13(8):921–928. https://doi.org/10.1007/s11869-020-00863-1

45. Singhal T (2020) A review of coronavirus disease-2019 (COVID-19). Indian J Pediatr 1–6:281. https://doi.org/10.1007/s12098-020-03263-6

46. Sohrabi C, Alsafi Z, O'Neill N, Khan M, Kerwan A, Al-Jabir A, Iosifidis C, Agha R (2020) World Health Organization declares global emergency: a review of the 2019 novel coronavirus (COVID-19). Int J Surg 76:71–76. https://doi.org/10.1016/j.ijsu.2020.02.034

47. Statista (2020) Estimated impact from the coronavirus (COVID-19) on India between April and June 2020, by sector GVA. https://www.statista.com/statistics/1107798/india-estimated-economic-impact-of-coronavirus-by-sector/

48. The Hindu (2021) https://www.thehindu.com/sci-tech/health/bharat-biotech-says-covid-19-vaccine-shows-81-interim-efficacy/article33980224.ece. Accessed on 02/04/2021

49. Torales J, O'Higgins M, Castaldelli-Maia JM, Ventriglio A (2020) The outbreak of COVID-19 coronavirus and its impact on global mental health. Int J Soc Psychiatry 0020764020915212:317. https://doi.org/10.1177/0020764020915212

50. Tripathi SN, Dey S, Tare V, Satheesh SK (2005) Aerosol black carbon radiative forcing at an industrial city in northern India. Geophys Res Lett 32(8). https://doi.org/10.1029/2005GL022515

51. Verelst F, Kuylen E, Beutels P (2020) Indications for healthcare surge capacity in European countries facing an exponential increase in coronavirus disease (COVID-19) cases, March 2020. Eurosurveillance 25(13):2000323. https://doi.org/10.2807/1560-7917.ES.2020.25.13.2000323

52. Vinjamuri KS, Mhawish A, Banerjee T, Sorek-Hamer M, Broday DM, Mall RK, Latif MT (2020) Vertical distribution of smoke aerosols over upper Indo-Gangetic Plain. Environ Pollut 257:113377. https://doi.org/10.1016/j.envpol.2019.113377

53. Wang J, Tang K, Feng K, Lv W (2020) High temperature and high humidity reduce the transmission of COVID-19. https://doi.org/10.2139/ssrn.3551767

54. Wang Z (2020) Studying the origin of COVID-19 from a systematic perspective. Preprints 2020, 2020040125. https://www.preprints.org/manuscript/202004.0125/v1
55. WHO (2020a) WHO characterizes COVID-19 as pandemic. 2020(3)
56. WHO (2020b) Coronavirus disease 2019 (COVID-19): situation report, 1–132
57. WHO (2020c) WHO coronavirus disease (COVID-19) dashboard. https://covid19.who.int/
58. WHO (2020d) Coronavirus symptoms. https://www.who.int/health-topics/coronavirus#tab=tab_3
59. WHO (2021) WHO coronavirus disease (COVID-19) dashboard. https://covid19.who.int/
60. World Air Quality Index Project (2021) https://aqicn.org/
61. World Bank (2021) World integrated trade solution: India trade. https://wits.worldbank.org/countrysnapshot/en/IND/textview
62. World Economic Forum (2020) World's 10 most polluted cities in India. https://www.who.int
63. WorldOMeter (2020) https://www.worldometers.info/coronavirus/
64. WorldOMeter (2021) https://www.worldometers.info/coronavirus/
65. Wu Y, Jing W, Liu J, Ma Q, Yuan J, Wang Y, Du M, Liu M (2020) Effects of temperature and humidity on the daily new cases and new deaths of COVID-19 in 166 countries. Sci Total Environ 729:139051. https://doi.org/10.1016/j.scitotenv.2020.139051
66. Xinhua, (2020). China's CDC detects a large number of new coronaviruses in the South China seafood market in Wuhan. Available at: https://www.xinhuanet.com/2020-01/27/c_1125504355.htm. Accessed 20 Feb 2020
67. Xu B, Gutierrez B, Mekaru S, Sewalk K, Goodwin L, Loskill A, Cohn EL, Hswen Y, Hill SC, Cobo MM, Zarebski AE (2020) Epidemiological data from the COVID-19 outbreak, real-time case information. Sci Data 7(1):1–6. https://doi.org/10.1038/s41597-020-0448-0

The Role of Innovation in the Health Crisis and the Sustainable Post-COVID Europe

Spatial Econometric Analysis of the Anti-COVID Measures Effectiveness and the Significance of the Luxury Industry in Shaping the Sustainable Future

Michał Taracha and Carmelo Balagtas

Abstract The multifaceted 2020 crisis, caused by the COVID-19 pandemic, affected both functioning of the economy and everyday life. The unprecedented character of the ways national economies had to deal with the COVID-19 outbreak had brought about the need for innovation development. At the same time, each country was affected by the effects of the pandemic to a different extent. A socioeconomic, spatial and cultural diversity of the European continent is a prime example illustrating different approaches in which countries attempted to deal with the disease, whose spread in Europe was diverse in terms of a time and degree to which each state was affected. This study aims to assess whether regional innovation was an important determinant of how well each European country dealt with the second wave of the pandemic. The second goal of this paper is to depict the role of innovation in shaping sustainable post-COVID future. In order to assess the effectiveness of anti-COVID measures and investigate the possible multifaceted causes of this effectiveness during the second wave of the pandemic (the period from 38 to 47th week of 2020 was selected), an OLS econometric linear model and ML spatial models were estimated. A number of excess deaths per 1000 inhabitants was chosen as a dependent variable. Observations consisted of 235 NUTS 2 regions of the European Union member states (without Austria and Ireland) and Switzerland. This econometric analysis of the current situation was enriched by the qualitative description revolving around the highly-innovative luxury and fashion

M. Taracha (✉)
SGH Warsaw School of Economics, Institute of Statistics and Demography,
al. Niepodległości 162, 02-554 Warsaw, Poland
e-mail: mtarac@sgh.waw.pl

C. Balagtas
SKEMA Business School, Sophia Antipolis Campus, 60 Rue Fedor Dostoïevski, 06902
Valbonne, France
e-mail: carmelo.balagtas@skema.edu

© The Author(s), under exclusive license to Springer Nature Singapore Pte Ltd. 2021   113
S. S. Muthu (ed.), *COVID-19*, Environmental Footprints and Eco-design
of Products and Processes, https://doi.org/10.1007/978-981-16-3856-5_5

industry, which was used as an example, concretizing how innovation contributed to the narrative of facing the crisis while developing more sustainable and long term solutions henceforth. An econometric analysis showed that a relatively high innovation level, including the quality of human resources (measuring the level of civil society) and intellectual assets, may mitigate the effects of the pandemic. Furthermore, the econometric analysis provided additional conclusions concerning other determinants of anti-COVID measures—behavior changes, psychological aspects, economic situation of modest households, lockdown stringency, testing rate, quality of the healthcare system, modern populist policies, percentage of people aged 65 years and more, and poor air quality. One of the most important findings provided by the SAR econometric model indicated that, among other factors, knowledge hubs can be important in interregional attempts to mitigate the effects of such challenges as COVID-19 pandemic. The second part of the chapter highlights the fact that the luxury industry, at its core, is sustainable and offers superior value and has an inherent ability to make timeless emotional connections in people's minds—which may prove to be important for the post-pandemic world. In the unprecedented turbulent times, more intimate values of luxury may prove to be more relevant than ever. The very meaning of luxury is also set to become much more diversified and contextual. The importance of the luxury industry in shaping the post-COVID sustainable future is associated with its ability to quickly adapt to the changing consumer expectations and societal landscape in general. The concept of revisiting its roots by the luxury industry concretizes the ideas behind the reality of our society's current state and the expectations thereafter due to global climate change.

Keywords Innovation · Excess mortality rate · Spatial autoregressive
(SAR) model · COVID-19 · Luxury and fashion industry · SHARE-COVID-19 ·
Populism · Spatial econometrics

1   Introduction

The COVID-19 pandemic has provoked a global socioeconomic disruption influencing various areas of human activity. The multifaceted crisis affected both functioning of the economy and everyday life, giving a stimulus to a wider (sometimes unprecedented) use of the latest technologies. Each country was affected by the effects of the pandemic to a different extent. Reasons for the differences in the intensity of COVID-19 effects were various, including the rigidity of national responses, population density, level of openness of each economy, the size of tourist flows, specificity of the culture and civil society related to the social development level, geopolitical localization, demographic situation or the history of each region. A socioeconomic, spatial and cultural diversity of the European continent is a prime example illustrating different approaches in which countries attempted to deal with the disease, whose spread in Europe was diverse in terms of a time and degree to which each state was affected. This is why Europe was selected to be a subject of an analysis that will depict some of the key determinants of anti-COVID measures effectiveness.

Since national capabilities, related to the economic development level, were crucial for the fast response and effective mitigation of pandemic repercussions (especially during the second wave of the disease), one of the most important determinants of how well European countries have managed to deal with the 2020 crisis is the level of regional innovation. The turbulent 2020 can be described as the year of rapid changes, that elevated the necessity for science and technology to combine its strengths and form solutions that may or may not have been thought of in the previous years. The unprecedented character of the ways national economies had to deal with the COVID-19 outbreak had brought about the need for innovation development. Thus, this paper will be also focused on how this development manifested the man's greatest capabilities to win over crises that comes his way and how the new solutions may be of a benefit for the future.

In order to assess the effectiveness of anti-COVID measures and investigate the possible multifaceted causes of this effectiveness, an OLS (ordinary least squares) econometric linear model and ML (maximum likelihood) spatial models were estimated. A number of excess deaths per 1000 inhabitants was chosen as a dependent variable—a measure of how well each country dealt with the crisis. The sources of the data included Eurostat, OECD Stat, European Centre for Disease Prevention and Control, SHARE, Regional Innovation Scoreboard, Our World in Data, Eurofound and World Health Organization databases. Observations consisted of 235 NUTS 2 regions of the European Union member states (without Austria and Ireland) and Switzerland. Furthermore, in order to depict future opportunities and circumstances that will shape the post-COVID world, the highly-innovative luxury and fashion industry was used a as key example, concretizing how innovation contributed to the narrative of facing the crisis while developing more sustainable and long term solutions henceforth.

The first segments of the chapter will be, thus, concentrated on factors which determined the anti-COVID policy effectiveness in Europe with the special focus on innovation aspect. A statistical description will be followed by the econometric analysis illustrating, among others, the relationships between regions in the studied countries. This econometric analysis of the current situation will be further enriched by the qualitative description revolving around luxury and fashion sector—depicting best practices that explain the ideas proposed in the earlier parts of the chapter while mentioning the examples of particular businesses and governments that have transformed their strategies in order to respond to the new unpredictable times with agility and efficiency.

2   COVID-19 Crisis in Europe

The COVID pandemic has provoked the biggest economic crisis since the Second
World War. Both the estimates from the World Bank and International Monetary
Fund predict that the 2020 economic recession will be more than twice as deep as
the decline brought about by the global financial crisis of 2007–2009.[1] The 2020
health crisis influenced various sectors from manufacturing, mining, quarrying, oil
extraction, transportation and warehousing—to educational, accommodational or
food services. McKinsey & Company Insights from 2020 show that the most
affected industries will get back to their level of contributions to the global GDP
after more than five years or much more in the case of small businesses that were hit
the most.[2] Since the emergence of global economy and the integration of national
economies into the global economic system, only world wars and the Great
Depression have prompted a comparable economic downturn and setbacks for
international trade. According to the International Monetary Fund projections, the
euro area production will return to its pre-crisis levels at the end of 2022.[3]

In terms of the number of cases and deaths, COVID-19 is the biggest pandemic
since the Spanish flu. Among other continents, Europe was profoundly stricken by
the pandemic with the number of reported deaths amounting to one third of the total
number of global deaths in December 2020. Meanwhile, the percentage of the
Europeans in the world population constitutes less than 10%. Only in the period of
January-November 2020, the number of deaths in the European Union exceeded the
multi-annual average pre-crisis level (for the same months) by approximately 385
thousand deaths, which gives an anomaly of additional 87 deaths per 100,000
inhabitants.[4]

At the same time, the extent to which the European Union has suffered from the
pandemic was not uniformly distributed across different countries or regions.
During the first wave, regions that suffered the most in terms of mortality anomaly
were located in Spain, Italy, France, Belgium and Netherlands. In particular, this
concerns regions of Castile, Catalonia, Navarra, Lombardy, Piedmont, Liguria,
Emilia-Romania, Île-de-France, Alsace, Limburg and North Brabant. The EU
countries that were stricken the most by the first part of the second wave (from the
38th to the 47th week of 2020) include Bulgaria, Romania, Poland, Czech
Republic, Spain (Aragon), Portugal (Alantejo), Italy (Northwestern part) and
Belgium (Wallonia). Mortality anomaly from two waves of the COVID pandemic
can be seen in Figs. 1 and 2.

---

[1]World Economic Outlook [65].

[2]COVID-19 recovery in hardest-hit sectors could take more than 5 years [17].

[3]Morcos [45].

[4]Ireland was excluded from these statistics.

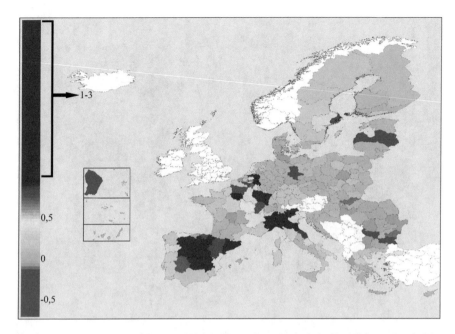

Fig. 1 Excess number of deaths per 1000 inhabitants from the period of late February–early May 2020 compared with the average level from 2014 to 2019 in the European Union (without Austria and Ireland) and Switzerland. *Sources* own elaboration based on the data of Eurostat and European Centre for Disease Prevention and Control

3 Innovation as a Driving Force to Deal with the COVID Crisis

One important factor for regional differences in the mortality level and the main potential contributor to the post-COVID recovery is innovation level—encompassing a high level of educated human resources, intellectual assets (for example, a high regional performance on trademark or design applications), decent R&D investment or employment in knowledge-intensive activities.[5]

Owing to new technologies, companies can decrease their dependency on the physical presence of their employees in the workplace. While technologies enabling large-scale remote work were already existing at the beginning of the pandemic, there is still a room for improvement of remote performance (decreasing the disconnect frequency, increasing the speed of file downloads and the connection quality).[6] An enormous shift to the remote work, which occurred during the first wave of 2020 pandemic, can be seen in the Fig. 4. In the European Union and in the

---

[5]Hollanders et al. [32], p. 6.
[6]O'Halloran [47].

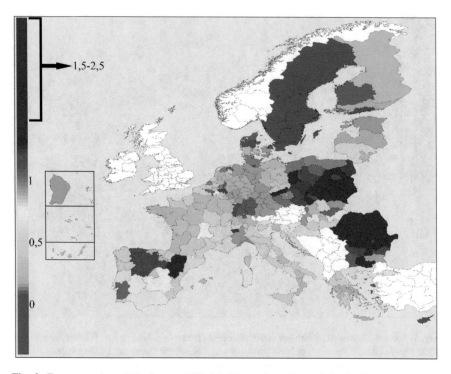

Fig. 2 Excess number of deaths per 1000 inhabitants from the period of mid-September–late November 2020 compared with the average level from 2014 to 2019 in the European Union (without Austria and Ireland) and Switzerland. *Sources* own elaboration based on the data of Eurostat and European Centre for Disease Prevention and Control

case of employees aged 50 years and more, however, it did not concern over a half of workers, despite the fact that the European Commission has reiterated the importance of telework in the context of the COVID-19 crisis.[7] This indicates that there is still a room for improvements in the area of remote work. New technologies are likely to influence especially those who are not familiar with business communication platforms and they constitute up to 43% of employees who started their work from home during the crisis, as this percentage declared to have learned new computer skills since the pandemic outbreak. The need for innovation in another area—of virtual education—concerns improvements in the possibility of performing fair live assignments reducing cheating, but also the possibility of collaboration with peers in breakout rooms. Furthermore, edtech investments can aim at enabling wide usage of active learning techniques and timely learner support in order to enhance learning effectiveness (Fig. 3).[8]

---

[7]Milasi et al. [44].

[8]Markova et al. [40], p. 686.

Percentage of people aged 50+ years from the European Union (without Austria and Ireland), Israel and Switzerland who, after the COVID-19 outbreak, worked:

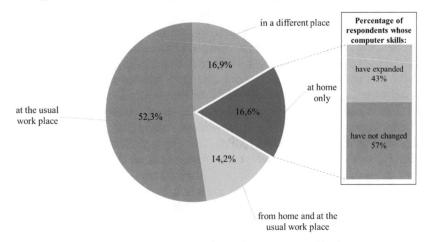

Fig. 3 Results of a SHARE survey conducted in June–August 2020 on 11,109 employed people aged 50 years and more and living in the European Union (without Austria and Ireland), Israel and Switzerland. *Source* own elaboration based on the data of SHARE database

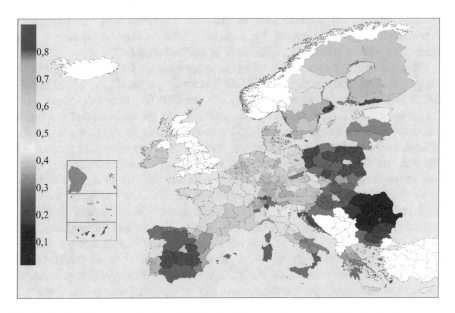

Fig. 4 Regional innovation index of 2019 in the European Union and Switzerland. *Source* own elaboration based on the data of Regional Innovation Scoreboard 2019 database

Innovation is also an important factor of the effectiveness of public policy dealing with the socio-economic impact of the 2020 crisis, and especially—the effectiveness of lockdown measures. In the European Union, since the beginning of the first pandemic wave, ten European governments have engaged in the development and implementation of anonymized phone location data. This technology can be used in tracing population movements to track the spread of the virus and to make the anti-COVID policy more informed. Moreover, it can be helpful in anticipating regional or national peaks of the pandemic which enables a more flexible adjusting of lockdown measures to the actual situation. At the end of March 2020, seven countries started initiatives aimed at the development of privacy-friendly contact tracing applications. These applications make it possible to alert those who voluntarily accepted to install them on their mobile phones when another owner of the app-equipped device, with whom they had gotten into contact with, got infected.This technology ensures a high speed of information about infected people and thus helps to effectively deal with coronavirus—which is spread mostly at the beginning of the infection.[9]

In the lockdown reality, new technologies flourished as well in the electronic-supported healthcare system practices, such as e-health, including health phone applications, referred to as mhealth. E-health technologies were being developed for over two decades before COVID-19 pandemic, however, they have been improved and appropriated after an introduction of national lockdowns. E-health applications support national healthcare systems with tailored digital consultation functions (granted by personal digital assistants), remote patient monitoring and by spreading health services and information.[10]

Other innovative technologies, which have been around for several years but are being perfected due to the SARS-CoV-2 spread, are used in the COVID-19 detection and treatment. A major example is the COVID vaccine developed with the messenger RNA (mRNA) technique which, unlike the conventional vaccine, is based on mRNA strings encoding spike proteins. Due to the cost-, time- and labor-consuming character of the conventional vaccines development process, the research on DNA vaccines began in the 1990s.[11] Another example is the pre-symptomatic and asymptomatic detection of the COVID-19 disease provided by the wearable monitoring devices. Wearable monitors for healthcare were first proposed 25 years before the COVID-19 outbreak. Nevertheless, traditional methods of disease detection are relatively expensive, slow and ineffective at early stages of disease. Similarly to the mRNA vaccine, wearable devices became effective to the point of being very profitable for the virus of SARS-CoV-2 parameters just a few years before the 2020 pandemic.[12] Other recent improvements include artificial intelligence enhancement of peptide-based therapeutics, which is

---

[9]Klonowska and Bindt [34], p. 9.

[10]Fejerskov and Fetterer [30], p. 12.

[11]Ball [5], p. 17.

[12]Bahmani [4].

developed since the 1990s.[13] Biotech companies, such as Nuritas, use artificial intelligence-based platform to discover antiviral peptides for COVID-19.[14]

Even in times when the pandemic is still in the full swing, numerous virologists and immunologists, such as Dan Barouch, argue that COVID-19 experience will reshape the vaccine science future.[15] However, the profound transformation provoked by the 2020 crisis will have a lasting impact on other industries as well—the most innovation-driven market sectors will be deeply transformed by the changes accelerated by the pandemic. In the face of the new market reality, marked with the digital escalation and rapidly changing consumer expectations, other specialists, including expert economists and marketers, such as Federica Levato (a co-author of Altagamma reports), argue that in 10 years from now, the profound transformations will change an already very innovative luxury industry into the market for "insurgent cultural and creative excellence".[16]

As engines of European economic development, regions with high innovation scores are likely to play a crucial role in the future economic recovery and in shaping the post-pandemic world. A number of studies concerning Regional Systems of Innovation identified several patterns in the regional innovation concentration. It tends to be irregularly distributed across regions and spatially concentrated over time. At the same time, regions with similar innovation scores tend to have a different economic growth patterns.[17] This phenomenon can be described by a tendency of regional innovation levels to be subject to the concentration and deconcentration processes stemming from geographic, economic, social and technical heterogeneity of the economic space and the resulting dynamics of economic development.[18] The uneven distribution of innovation level and the geographical concentration of regions with similar innovation scores is illustrated in Fig. 4.

4   Other Contributors to Cross-Regional Differences in Effectiveness of Anti-COVID Policy

Separate components of regional innovation, such as human resources quality, can be also decisive for the lockdown effectiveness. A number of studies find empirical and theoretical evidence for the positive relationship between innovation, human capital and social capital.[19] The substance of social capital lies in the sympathy, trust, and forgiveness offered to a person by other people. Thus innovation is also

---

[13]Cf. Basith et al. [6], pp. 1276–1314.

[14]Burke [10].

[15]Ball [5], p. 16.

[16]D'Arpizio and Levato [20].

[17]See Footnote 5.

[18]Kuciński [36], pp. 19–22.

[19]Cf. Dakhli and De Clercq [19], pp.107–128.

strongly positively correlated with the 'goodwill' offered to other people.[20] Caring for other individuals, associated with the high quality of human resources, keeps people wearing face masks (which are much more effective in preventing the virus from spreading when only the potential spreader wears a mask than in the situation when only the receiver wears it) even when it is uncomfortable to the wearer. The possible presence of this kind of 'goodwill' occurs in the society because of the high robustness of reciprocity among humans which co-occurs in wider social networks.[21] This robustness indicates that social distancing of younger adults (who have a relatively high COVID-19 basic reproductive rate) may be enhanced by the desire to protect older members of their families.

Difficulty of social distancing is also associated with the relationship orientation embedded in the culture. Some theoretical approaches to culture differences make the synthesis of certain culture dimensions, along which cultural values can be analyzed. The relationship orientation, derived from the synthesis of Hofstede's individualism-collectivism and power distance dimensions and Hall's high-low context dimension, is sometimes opposed to the information orientation. In this understanding, relationship-oriented cultures tend to focus more on relationships and reduction of transaction costs rather than on information and competition. Thus, in some countries whose cultures are more relationship-oriented, social relationships are fundamental for the interpretation of symbols and signs, while in the more information-oriented countries, information and competition is more salient. Relationship-orientation tends to go in line with collectivism, vertical organization, high power distance and language indirectness.[22] Other approaches, such as the one made by an application of Aperian Global consulting company, juxtapose relationship orientation and task orientation in one dimension. On this continuum, more relationship-oriented cultures tend to prioritize maintaining personal relationships in the organization over focusing on task accomplishment.[23] Both approaches imply that countries with more relationship-oriented cultures can deal with social distancing and productivity maintenance during lockdowns less effectively.

Considering that at the beginning of the pandemic more open European economies were potentially more vulnerable to the virus gaining a foothold in their territory, one could expect that economic openness makes it more difficult to keep control over the spread of the virus. Nevertheless, specialists argue that isolationist economic policies might hamper the ability to face COVID-19 with their negative impact on health sector.[24] An analysis of regional economic openness of 209 European NUTS 2 regions shows that during the first pandemic wave (from 9 to 19th week of 2020) a correlation between excess mortality and regional exports divided by regional GDP is negative—it is equal to -0.051. The analogical

---

[20]Adler and Seok-Woo [2], p. 18.

[21]Melamed et al. [43], p. 6.

[22]Chandler and Graham [11], pp. 4–5.

[23]Understanding Your GlobeSmart Profile [63].

[24]Clausing [12].

correlation for the first part of the second wave (from 38 to 47th week of 2020) amounts to −0.207. In the case of weighted regional export scores (encompassing export value and the number of export destinations) divided by regional GDP, the analogical correlations for the first and second wave were of −0.116 and −0.204.[25] Furthermore, bearing in mind the visualizations of excess moralities that were presented in Figs. 1 and 2, it seems that the regional GDP per capita or urbanization rate can be a contributor to the anti-COVID policy effectiveness. An analysis of 244 NUTS 2 regions shows that the correlation between the regional GDP per capita and regional excess mortality was of 0.170 for the first wave and of −0.421—for the second wave. In the case of urbanization rate, the respective correlations with excess mortality were of 0.183 and −0.126.[26] These statistics show that more economically open, wealthier and urbanized European regions were likely to perform relatively better when the pandemic was advancing in time.

What is more, the health crisis and associated lockdowns had a great psychological impact, affecting especially elderly, having an increased risk of being severely affected by the virus. The COVID-19 outbreak is officially recognized as the pandemic since March 2020, but it has already taken a dramatic toll on everyday life. According to the SHARE COVID-19 Survey (conducted between June and August 2020 on the group of more than 52,000 respondents aged 50 years or more), 30.3% of people questioned stated that they have felt nervous, anxious, or on edge in the month prior to the questionnaire. As much as 70.9% out of this 30.3% admitted to feel that way "more so" than before the Corona outbreak. Similarly, out of 26% of people surveyed declaring that they felt depressed, 63.2% expressed that it was due to coronavirus. Almost one third of respondents indicated that they felt lonely in the previous month and more than 40.3% of them pointed out that COVID was the factor for their unpleasant emotional state.[27] It is possible that the anti-COVID policy is more effective in countries whose citizens were less affected psychologically by the crisis. More contented society is likely to be more enduring in adhering to lockdown restrictions.

Likewise, another determinant of how well regions cope with the COVID crisis is related to restrictions. An Oxford COVID-19 Government Response Tracker provides a comparable measure of government responses to the pandemic. It shows that there is a negative correlation between the stringency of restrictions and economic activity.[28] According to Our World in Data[29] and ECDC[30] statistics, government responses were different across different European countries. In the 10th

---

[25]Own elaboration based on the data concerning exports strength obtained from Eurostat and PBL Netherlands Environment Assessment Agency [27].

[26]Own elaboration based on the data obtained from Eurostat, Statista.com and Swiss Federal Statistic Office.

[27]Own elaboration based on the data obtained from SHARE database.

[28]Directorate-General for Economic and Financial Affairs of the European Commission [26], pp. 14–15.

[29]COVID-19: Stringency Index [18].

[30]Data on the weekly subnational 14-day notification rate of new COVID-19 cases [23].

week of 2020, the stringency index in Czech Republic was of the second highest value in Europe after Italy (stringency indexes in these countries were of 16.7 and 19.4 respectively), having more than 230 times greater number of active COVID-19 cases than Czech Republic. In the same week, there were no restrictions counting to the stringency index of the Netherlands, despite the fact that the reported number of active cases in this country was eight times bigger than in Czech Republic. In the 13th week of 2020, Spain was almost catching up with Italy in terms of the number of cumulative COVID cases (with respectively 96 and 97 thousand reported cases). Nevertheless, the stringency index of Spain was almost 3 times lower than in Italy. At the same time, stringency in Poland was almost 2.3 times more strict than the one in Sweden even though the number of reported cases in Sweden was 2.1 times greater than in Poland. The situation was similar for France and Germany—French restrictions exceeded strictness of German stringency by 52%, while the number of cases was larger in Germany by 43%.

Government reaction can be also reflected by the number of tests related to the number of population and an associated measure—testing positivity rate (on average, the lower the testing rate, the higher the testing positivity rate), which is presented in Fig. 5. According to Centers for Disease Control and Prevention recommendations, viral tests of both symptomatic and asymptomatic individuals are important,[31] as SARS-CoV-2 specificity makes it impossible to rely solely on symptom-based contact tracing, since up to 45% of those infected can be asymptomatic[32] and approximately 50% of transmissions occur during early phases (also pre-symptomatic) of the COVID-19 infection.[33] The data of the last "Release 0.0.1 beta of SHARE wave 8 COVID-19" shows that behavioral changes of economic agents were also different across Europe. Figure 6 illustrates the frequency of wearing face masks outside of home by people aged 50 years and more in the European Union (except Ireland and Austria) and Switzerland from June 3 to August 14. To enable a more reliable visualization, data on some regions from respondents, were aggregated at a level higher than NUTS 2. When preparing the map, a fairly strong assumption was made, according to which a variable expressed on an ordinal scale was treated as a variable on an interval scale—the arithmetic mean of its individual values was calculated after prior ranking. According to this assumption, subsequent answers concerning the frequency of wearing the mask outside of home ("always", "often", "sometimes", "never") were assigned successive natural numbers (equally distant from each other). In other words, it is assumed that the differences between adjacent responses are the same.

Since government restriction policies and behavioral changes constitute a relevant factor for how well European economies deal with the pandemic, one can also consider the role of populism and its tendency of restricting civil society.[34] Taking

---

[31]Overview of Testing for SARS-CoV-2 (COVID-19) [48].

[32]Cf. McAdam et al. [42].

[33]See Footnote 9.

[34]Liddiard [38].

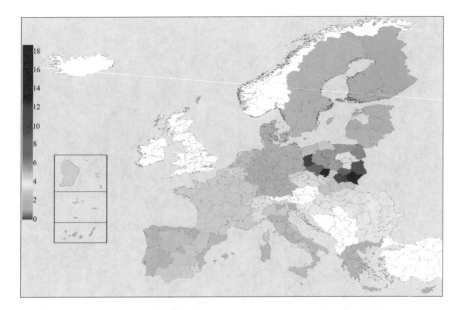

Fig. 5 Testing positivity rate in the 45th week of 2020. *Source* own elaboration based on the data of European Center for Disease Prevention and Control

the approach of Cas Mudde, populist ideology divides the electorates into two antagonistic groups of "the pure people" and "the corrupt elite".[35] Growing parliamentary influence of right-wing populist parties and associated anti-establishment voter share pose a threat of different extent on democratic durability in the whole Europe.[36] Since the pandemic has increased economic uncertainty,[37] the inevitable rise of economic insecurity (during and after the 2020 crisis) is likely to trigger the further increase in anti-establishment popularity connected with populism.[38] This is the reason why right-wing populism might be a significant contributor to the country's effectiveness of dealing with the COVID-19 pandemic.

Firstly, populism is related to the increase in size of the state-controlled sector which, for many experts, is considered to be more costly and inefficient; which is also amplified by the noticeable transition to a more left-leaning socio-economic profile of right-wing populists.[39] Countries of Central and Eastern Europe seem to be especially prone to the excessive bloating of the government, as in some countries of this region, national populists do not need to enter in coalitions with

---

[35]Ucen [62], pp. 49–62.

[36]Timbro Authoritarian Populism Index [60].

[37]Directorate-General for Economic and Financial Affairs of the European Commission [26], p. 18.

[38]See Footnote 35.

[39]Röth et al. [56], pp. 327–328.

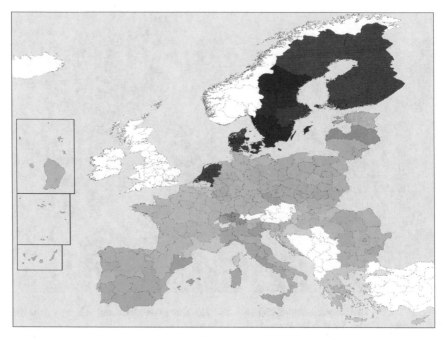

Fig. 6 Frequency of wearing face masks outside of home by people aged 50 years and more (dark green—"always wearing a mask when going outside", dark red—"never wearing a mask when going outside"). *Source* own elaboration based on the data of SHARE database

other right-wing parties and make concessions liberalizing their economic policy. Secondly, populism is associated with the denial of a pluralistic political system and restraint of civil society. Civil society can be understood as a measure making some societies able to cope more efficiently than others with difficult socio-economic and political, but also environmental challenges, and thus it is decisive in coping with 2020 crisis.[40] Furthermore, even if they are not completely anti-democratic, national populists tend to display an authoritarian-style bending of constitutionally provided rules in the country to their advantage.[41] This can be associated with the more inefficient anti-COVID policy which first and foremost takes advantage from the pandemic to repress political opponents and maintain power by undermining legislative power, profiting from a drop in public protests (which was of 60% when the first wave of the pandemic was in a full swing[42]), at the expense of taking effective measures to face the pandemic. It was the case in Poland where municipalities governed by the authorities associated with the ruling populist party (accounting for 9% of all municipalities) received over 10 times more funds per

---

[40]See Footnote 34.

[41]Ucen [62], p. 53.

[42]Pinckney and Rivers [49], p. 24.

capita from the Government Fund for Local Investments (RFIL) than other municipalities in January 2021.[43]

Some right-wing populist governments try to benefit from the pandemic to undermine the role of civil society in environmental protection. It was the case in Slovenia in April 2020, when the newly elected government announced an Act Amending the Act Determining the Intervention Measures to Contain the COVID-19 Epidemic and Mitigate its Consequences for Citizens and the Economy, comprising an article 42 that excluded civil society and scientific experts from participation in the construction of large construction projects,[44] violating both the UNECE Aarhus Convention and Slovene Constitution.[45] Similarly, according to the Polish Ecological Youth Forum, in July 2020, the populist government of Poland accepted the Draft Act Amending Certain Acts Supporting the Development of Housing allowing for partial privatization of land belonging to the publicly owned State Forests under the guise of the implementation of anti-COVID regulations. The public forest area planned for such unlawful sale is equal to nearly 1% of an entire area of the country.[46]

This is related to another determinant of how well regions deal with the COVID-19 pandemic—air pollution. An exposure to air pollution increases the severity of respiratory system diseases, such as COVID-19. According to the assessments of the cardiovascular experts, an exposure to particulate air pollution is responsible for a 15% increase in COVID-19 mortality worldwide and 19% in Europe.[47] Lower air pollution is also indirectly connected with the possibility of low carbon economic recovery from the crisis, which enhanced the role of the state as an economic decision-maker. Low carbon energy system development is a contributor to the more efficient anti-COVID policy because the system based on renewable energy is more resilient than the fossil fuel industry in times of economic crises.[48] In addition to its resiliency, energy sectors relying on renewable energy on average employ more people than their fossil fuel counterparts.[49]

---

[43]DGP: RFIL rozdzielony według klucza politycznego [25].

[44]Zakon o spremembah in dopolnitvah Zakona o interventnih ukrepih za zajezitev epidemije COVID-19 in omilitev njenih posledic za državljane in gospodarstvo (ZIUZEOP-A) [67], p. 2273.

[45]Exposed: Slovenian government exploits pandemic to shatter environmental safeguards [28].

[46]Skok na polskie lasy!? [58].

[47]Pozzer et al. [51], p. 2251.

[48]Blondeel [8], pp. 4–6.

[49]Policy Brief 13: Interlinkages Between Energy and Jobs [50] p. 2.

5 Linear Econometric Models of Excess Mortality Levels During COVID-19 Pandemic

In order to empirically determine relevance of aforementioned factors of the regional mortality rate anomaly linked to the COVID-19 pandemic and to obtain the information about the significance of the regional innovation index among other determinants, three econometric models, based on cross-sectional data, were prepared with an endogenous variable illustrating the number of excess deaths per 100,000 inhabitants of 2020, compared with the average level from 2014 to 2019. Each variable consists of 235 observations—NUTS 2 regions of the European Union (without Austria and Ireland) and Switzerland.[50]

Dependent variable in Models 1 and 2 represents the excess mortality of the second pandemic wave (from 38 to 47th week of the year) and the one in Model 3 shows the excess mortality of the end of the first wave (from 18 to 24th week of the year). The data to calculate both endogenous variables was obtained from Eurostat.[51] Due to a lack of the most recent values from weeks 44 to 47 for four countries, the missing values were imputed using the information about the regional number of reported COVID-19 cases from European Center for Disease Prevention and Control.[52] The missing values encompassed three last weeks for Greek and Italian regions and four weeks for Romania and Slovakia. The imputation was based on the modified Holt's two-parameter model of exponential smoothing with the trend component multiplied by a number reflecting the COVID-19 lethality. This number was set in the process of maximization of the resemblance between the modified variable presenting the number of reported infection cases (moved back by one week to reflect the median time to COVID-19 death) and the variable concerning the excess mortality. In this way the further extrapolation (based on the modified variable) was stable. Due to the fact that the lethality decreases over time, the imputed data was subsequently expressed as the number of deaths and aggregated at the national level, and then multiplied by another constant taking into account the real data about the number of COVID-19 deaths of the country which was available at the national level. In this way, values at the regional level were obtained—they were adding up to the new value taken into account the national number of COVID-19 deaths, but distributed among different regions using the same coefficient as before the aggregation.

The first exogenous variable, *RIS*, originates from the 2019 Regional Innovation Scoreboard and reflects the level of regional innovation at the level of NUTS 2.[53]

---

[50]Own elaboration. Variables of econometric models where constructed based on a data from Eurostat, OECD Stat, European Centre for Disease Prevention and Control, SHARE, Regional Innovation Scoreboard, Our World in Data, Eurofound and World Health Organization databases. See References [51–56], [58–61] and [63].

[51]Deaths by week, sex and NUTS 2 region [24].

[52]Data on the weekly subnational 14-day notification rate of new COVID-19 cases [23].

[53]Regional Innovation Scoreboard 2019 Interactive Tool [55].

HumanResources—the variable reflecting human resources quality, one of 18 component indicators (each component indicator is made of several more components) of the Regional Innovation Scoreboard, was added separately to Models 1 and 2. Model 3 consists of another component indicator *IntellectualAssets*—measuring intellectual assets performance.

The next set of variables—*RelationshipOrientation*, *LongTermOrient* and *LeftHome*—presents culture patterns and behavioral changes of communities in analyzed regions. First two variables are differentiated at the national level. *RelationshipOrientation* was retrieved from the map showing relationship-task cultural dimension illustrated in the Country Navigator User Guide presentation.[54] *LongTermOrient* reflects the long-term orientation of country's culture describing the strengths of links maintained by the society with its past when dealing with the present challenges, such as 2020 crisis. It was obtained from the Hofsede Insights country comparison.[55] The values for Greece were also assigned to Cyprus, for which the value of long-term orientation was missing. *LeftHome* reflects the percentage of people aged 50 years and more who left their home from the pandemic outbreak to the period of June–August 2020. It originates from the "Release 0.0.1 beta of SHARE wave 8 COVID-19".

Another variable obtained from SHARE database is *DipIntoSavings*. It shows the percentage of the region's population that had to dip into savings to cover the necessary regular daily expenses since the COVID-19 outbreak. It was added to reflect the economic situation of an average household in the region and it later turned out to be more significant than regional GDP per capita expressed in purchasing power standard units which was excluded from the model.

Variables *HappinessJuneJuly*, *LifeSatisJuneJuly*, *LonelyChange* and *AnxietyRateCOVID* illustrate the psychological state of each region's society. First three of them were obtained from the Eurofound survey "Living, working and COVID-19"[56] and they are differentiated at the national level. *HappinessJuneJuly* and *LifeSatisJuneJuly* express respectively the country's happiness and life satisfaction in the period of June–July 2020. These two variables were kept in the model despite the risk of collinearity (variance inflation factors of both variables slightly exceeded 10) because individual exclusion of each of them from the model does not change the statistical significance of another one and values of their coefficients do not entirely cancel each other out when added up (value of *LifeSatisJuneJuly* coefficient being approximately two times stronger despite the same scale used to present both of them). They were kept in the model considering that happiness can reflect recent psychological state of respondents while life satisfaction might show a more long-term one. Bearing in mind the 'Candril Ladder'[57] method of reporting life satisfaction, it is also more subjective than happiness. *LonelyChange* reflects the difference among

---

[54]Country Navigator—your tool for global skills development [15].
[55]Country Comparison [14].
[56]Quality of life and quality of society during COVID-19 [53].
[57]Cf. Mazur et al. [41], pp. 182–189.

percentages of respondents declaring that they have felt lonely from periods of April–May (comprising the peak of the first wave) and June-July (comprising the period of appeasement of lockdowns in Europe). *AnxietyRateCOVID* is differentiated at the regional level and represents the percentage of SHARE respondents who stated that they had felt nervous, anxious, or on edge in the month prior to the questionnaire conducted between June and August 2020.

Variables *HealthcareTrustJuneJuly*, *HealthcareTrustChange*, *StringencyWeeks27–43*, *PositivityCountry1W*, *PositivityRegion2W* and *TestingRegionEnd2W* concern the quality of healthcare system and government policy perceived by the society. First two variables originate from the Eurofound survey and show public trust in the healthcare system in the period of June-July 2020 and the difference of this trust between April–May and June–July 2020. *StringencyWeeks27–43* illustrates the value of Government Response Stringency Index published by Oxford COVID-19 Government Response Tracker of Blavatnik School of Government.[58] *PositivityCountry1W* reflects the average testing positivity rate from 18 to 26th week of 2020 and was calculated owing to the ECDC database.[59] Unlike previous variables, differentiated at the national level, *PositivityRegion2W* and *TestingRegionEnd2W* show the values of testing positivity rate and testing rate for each NUTS 2 region, with several exceptions depending on the ECDC data availability.[60] *PositivityRegion2W* concern the 45th week, whereas *TestingRegionEnd2W*—weeks 50–52 of 2020.

GovTrustChange, *MediaTrustChange* and *PressFreedom* originate from Eurofound survey and Reporters Without Borders' data concerning 2020 World Press Freedom Index.[61] These variables were added in order to reflect the populism severity at the country level, as controlling the media is a key objective of populist governments compared with other authorities. Moreover, populist rule displays a specifically negative relationship with all measures of press freedom, which tends to exacerbate if an economic profile of populist government is left-leaning.[62]

Another predictor, *PM10*, was added to the Model 3 with an intention to reflect the impact of the poor air quality linked to the particulate air pollution. This variable shows the average concentration of PM-10 particles (measured in $\mu g/m^3$) in the NUTS 2 regions. Its values were obtained after processing the data from the World Health Organization database—the data were initially averaged from all available years (annual average values for various years from 2013 to 2017 were available) at the level of a single measurement station, and then at the regional level. Its addition to the model was also dictated by the fact that the average concentration of PM-10 particles in the air weighted by population density is a component indicator of the

---

[58]COVID-19 Government Response Tracker [16].

[59]Data on testing for COVID-19 by week and country [22].

[60]Data for the maps in support of the Council Recommendation on a coordinated approach to the restriction of free movement in response to the COVID-19 pandemic in the EU/EEA [21].

[61]World Press Freedom Index [66].

[62]Kenny [33], pp. 261–275.

Better Life Index measure, the design of which takes into account the recommendations of the Stiglitz Commission. This indicator represents overall air pollution as a component of this Index, developed by Organization for Economic Cooperation and Development.[63]

The remaining variables *Percentage65+*, *sq_Density2019* and *ExcessMortWeeks9–23* were obtained from Eurostat and reflect the percentage of people aged 65 years and more in each region, square of the region's population density in 2019, and excess mortality in weeks 9–23 of 2020 calculated in the same way as the endogenous variable of Model 1.

Values of coefficients of explanatory variables and *p*-values of T-tests (showing their significance), along with model properties tests of Models 1–3 are presented below (Table 1).

The RESET test results for all three models showed that there is no ground to reject the null hypothesis about the correctness of the linear form of the model at 1% significance level (all *p*-values being equal to at least almost 0.05). Similarly, White test results indicated no statistically significant evidence for heteroscedasticity of an error term. At the same significance level, one can draw a conclusion about the lack of normal distribution of residuals only in the case of Model 3. However, this result is of a lower importance in the case of models with as much as 235 observations. Furthermore, coefficients in all models are statistically significant at the significance level of 5%—*p*-value in the T-test of only one coefficient, of *ExcessMortWeeks9–23* variable, slightly exceeds this level. R^2 ratios in Models 1–3 amounted to respectively 0.806, 0.809 and 0.257.

Regarding Model 1, the sign of the coefficient value of the *RIS* variable is negative. This indicates that, *ceteris paribus*, higher regional innovation score is associated with a lower mortality anomaly in the second wave of the pandemic. The value of the coefficient amounts to −35.634. It shows that, all other things being equal, an increase of innovation index in one region by 0.1 (a close equivalent of the difference between Madrid and Castile and León region or between Brussels and Wallonia, or between Piedmont and Apulia located in opposite sides of the Italian Peninsula) goes in line with a decrease of excess mortality by 3.6 deaths per 100,000 inhabitants in the same region. For the *HumanResources* variable, the impact of such unitary change on mortality level is even greater—an increase of Human Resources Indicator by 0.1 (equal to a difference between Germany and Czech Republic or between Poland and Hungary) is associated with 2020 mortality anomaly lower by 9.6 deaths per 100,000 people.

A coefficient of *RelationshipOrientation* variable indicates that, *ceteris paribus*, a difference between the most relationship-oriented Portugal and the most task-oriented Finland is responsible for approximately 149.6 additional deaths per 100,000 inhabitants in the second wave of 2020 pandemic. The value of coefficient of *LeftHome* variable shows that the percentage of elderly people not leaving their home during, approximately, first 22 weeks of the COVID-19 pandemic higher by

---

[63]Compendium of OECD Well-being Indicators [13], pp. 12–34.

Table 1 Results of model properties tests for Models 1–3

Test	H0	p-value		
		Model 1	Model 2	Model 3
RESET test	Regression is linear	0.047	0.046	0.197
White test	Error term is homoscedastic	0.102	0.042	0.018
Test of normal distribution of residuals	Normal distribution	0.138	0.098	0.002

Source own elaboration

Model 1 The linear OLS model explaining excess death number per 100,000 inhabitants in 2020 (weeks 38–47)

Variable	Coefficient	p-value
const	−269.870	0.0001
RIS	−35.634	0.0337
HumanResources	−116.283	9.81E-17
RelationshipOrientation	0.546	3.76E-24
LeftHome	45.900	0.0078
DipIntoSavings	217.402	1.02E-12
HappinessJuneJuly	−51.115	0.0008
LifeSatisJuneJuly	101.067	1.17E-11
LonelyChange	7.535	6.89E-13
AnxietyRateCOVID	45.843	0.0041
HealthcareTrustJuneJuly	−15.514	4.47E-06
StringencyWeeks27–43	−3.303	8.20E-17
PositivityRegion2W	0.615	2.19E-08
TestingRegionEnd2W	−0.004	0.0002
GovTrustChange	20.872	0.0019
MediaTrustChange	35.252	0.0028
PressFreedom	−2.498	0.0001
Percentage65+	263.589	3.83E-08

Source own elaboration

1% point is related to, *ceteris paribus*, the mortality anomaly lower by 49.5 deaths per 100,000 people.

DipIntoSavings variable depicts that a deterioration of economic situation of modest households associated with the 2020 crisis makes things worse as each additional percentage point of people who had to dip into their savings due to the pandemic is responsible, all other things being equal, for an augmentation of death rate anomaly by as much as 2.2 deaths per 100 inhabitants.

Since the lack of collinearity can be questioned for the two next variables *HappinessJuneJuly* and *LifeSatisJuneJuly*, one should be careful to draw conclusions from values of their coefficients. An attempt to remove variables from the model indicates that the less significant variable *HappinessJuneJuly* might carry a

Model 2 The linear OLS model explaining excess death number per 100,000 inhabitants in 2020 (weeks 38–47)

Variable	Coefficient	p-value
const	−233.055	0.0009
RIS	−43.333	0.0116
HumanResources	−113.929	2.97E-16
RelationshipOrientation	0.536	1.60E-23
LeftHome	49.539	0.0041
DipIntoSavings	212.425	2.64E-12
HappinessJuneJuly	−56.326	0.0002
LifeSatisJuneJuly	102.606	5.00E-12
LonelyChange	7.149	1.30E-11
AnxietyRateCOVID	43.650	0.0060
HealthcareTrustJuneJuly	−15.725	2.96E-06
StringencyWeeks27–43	−3.386	1.88E-17
PositivityRegion2W	0.598	4.55E-08
TestingRegionEnd2W	−0.004	0.0003
GovTrustChange	19.138	0.0045
MediaTrustChange	36.627	0.0018
PressFreedom	−2.608	0.0001
Percentage65+	255.119	9.16E-08
ExcessMortWeeks9–23	7363.120	0.0516

Source own elaboration

Model 3 The linear OLS model explaining excess death number per 100,000 inhabitants in 2020 (weeks 18–24)

Variable	Coefficient	p-value
const	15.644	6.89E-05
PM10	0.201	0.019
IntellectualAssets	−8.645	0.0282
LongTermOrient	−0.091	0.0025
HealthcareTrustChange	10.525	9.9E-07
PositivityCountry1W	1.247	2.46E-09
PressFreedom	−0.478	8.16E-06
sq_Density2019	−1.77E-07	0.0425

Source own elaboration

slightly disturbing amount of redundant information, but the negative sign of the coefficient of *LifeSatisJuneJuly* is common in various variants of the model. This negative relationship between life satisfaction and the effectiveness of anti-COVID policy can be related to the weaker orientation towards change and a stronger orientation towards stability of individuals with a high life satisfaction. Thus, according to self-regulatory theory, people that are more satisfied with their lives tend to be more prevention-focused which is associated with avoidance of behaviors that might change their current situation.[64] In this understanding, it might be easier for less satisfied and more promotion-oriented people to accommodate to lockdown reality and submit to the new rules.

The value of coefficient of *LonelyChange* variable shows a tradeoff between loneliness and anti-COVID measures effectiveness. Bearing in mind that this variable compares two periods and the later period (of June–July) was associated with lifted restrictions, higher loneliness during the first wave of pandemic, compared to the situation with much more re-opened economy and free movement, goes in line with the lower excess mortality rate during the second COVID-19 wave. To understand this conclusion, one should consider the fact that the *LonelyChange* variable measures the difference between percentages of lonely individuals during lockdown and those after lockdown in percentage points. Each percentage point by which the percentage of lonely people during the lockdown is higher (compared to the percentage of lonely individuals during a more regular situation—similar to the pre-crisis one) decreases the excess mortality by 7.5 deaths per 100,000 people.

No such tradeoff can be observed for the next variable, *AnxietyRateCOVID*. On the other hand, an increase by one percentage point of people who feel nervous, anxious or on the edge due to COVID-19 (among all individuals feeling this way) is associated with 45.8 additional deaths per 100,000 inhabitants of a certain region. *HealthcareTrustJuneJuly* shows that the more trust society placed in the healthcare system, the higher its effectiveness during the 2020 crisis was.

The conclusions coming from the first model confirm the importance of other measures on which the government has a more direct influence, such as strictness of lockdown and testing rate. The coefficient value of *StringencyWeeks27–43* variable illustrates that, *ceteris paribus*, a one-point show that a higher testing rate also goes in line with the lower excess mortality rate.

The next set of variables—*GovTrustChange*, *MediaTrustChange* and *PressFreedom*—show the quality of a general public policy and a degree of populism. Coefficient of *GovTrustChange* allows for a conclusion that each percentage point by which the level of trust towards the government was lower during the crisis, compared to the more usual situation, increases the number of deaths per 100,000 inhabitants by 20.9 (all other things being equal). The impact of trust towards national news media is the same in its negative relationship with mortality anomaly and it is even by 60% stronger than the impact of the change of how the government is trusted. *PressFreedom* indicates that a one-point increase of the Press Freedom Index (expressed on a scale from 1 to 100) is related to a decrease of excess mortality by 2.5 deaths per 100,000 people. This result is unintuitive as values of Press Freedom Index are higher for countries having more restrained media independence. Nevertheless, considering the conclusions associated with other variables, the last result does not detract from the importance of civil society and a more democratic-style of exercising power, it rather shows that less plural media do not necessarily work to a detriment of the society's common goal to

---

[64]Luhmann and Hennecke [39], pp. 54–56.

mitigate the effects of the pandemic during COVID-19 crisis, even if staying in power is a main goal of populist authorities restraining this media. Furthermore, the correlation of both variables *GovTrustChange* and *PressFreedom* with the variable reflecting the percentage of anti-establishment voters in a country is positive and similar (of 0.18 and 0.20). The correlation between *MediaTrustChange* and share of anti-establishment voters is negative but it is related to the fact that the trust towards media in countries staying under populist rule was already very low, and thus less likely to further decrease. It is better described by the levels (and not the change) of trust towards government and media after the lockdown. These levels were negatively correlated with the percentage of anti-establishment voters and, in the case of trust towards news media, this correlation was even stronger than the one of *PressFreedom*, and amounted to −0.34. To sum up, not the restrained media freedom (reflected by the Press Freedom Index) but the poor media quality (reflected by the changing level of trust) is a more harmful feature of populist authorities during COVID-19 crisis. Nevertheless, these two features are connected as in some European countries with populist rule only 25% of all information broadcasted in the public television is non-manipulative and of a purely informative character.[65] Moreover, populist governments are also more likely to lose general public trust during the pandemic—possibly because of the abusing policies described in the previous section.

The coefficient value of the last variable of Model 1, *Percentage65+*, indicates that a 1-percentage-point increase in the percentage of people aged 65 years and more in a region's society goes in line with a rise of the 2020 excess mortality rate in this region by 2.6 deaths per 1000 people.

Model 2 differs from Model 1 in that it includes *ExcessMortWeeks9-23*. This additional variable shows that regions that had a higher excess mortality in the first pandemic wave have, *ceteris paribus*, a higher excess mortality during the second wave. This indicates that the herd immunity threshold, whose estimates from December 2020 are even of 90%,[66] was not of importance in 2020 at the regional level.

Model 3 was prepared to investigate the relevance of air pollution during the end of the first wave of COVID-19 pandemic. This model was needed as the *PM10* variable constituted a statistically insignificant predictor of 2020 mortality anomaly for the first wave as a whole and for the second wave, even though its coefficient value had an intuitive positive sign in both cases. *PM10* shows that, all other things being equal, an increase of 10 $\mu g/m^3$ of PM10 particles is responsible for a growth of 2 excess deaths per 100,000 inhabitants during the period ranging from the 9th to 19th week of 2020. One relevant finding connected with *PM10* insignificance in the models explaining different phases of the pandemic is that a poor air quality may extend the duration of the pandemic wave after its peak is reached—particulate air

---

[65]Sprawozdanie ze stanu ochrony języka polskiego za lata 2016–2017. Język informacji politycznej [59].

[66]Allen [3].

pollution is likely to prolong the fading phase of the pandemic or make it more severe.

Coefficients of other variables of Model 3 indicate that a high performance of intellectual assets and a relatively long-term oriented culture are associated with a lower mortality anomaly connected with the end of the first pandemic wave. Another additional variable, *HealthcareTrustChange*, indicates that if a public trust towards the healthcare system during the first wave of pandemic, compared to the level of trust from the time when there were less restrictions, is larger, the mortality rate is reduced. The only variable providing less intuitive conclusions is *sq_Density2019*. It shows that there is a level of population density above which an excess mortality during the end of the first COVID-19 wave was decreasing. It may be associated with the low R^2 ratio and the fact that this variable captured the variability of other factors related to large cities, which were not included in the model. Examples of such factors can be related to better living conditions or a higher percentage of people wearing masks in larger cities.

6 Spatial Econometric Models of Excess Mortality Levels During COVID-19 Pandemic

In order to visualize the impact of the unitary change in the selected variable in one European region on the excess mortality level (reflecting anti-COVID policy effectiveness) of different regions, the subsequent spatial econometric models were constructed using the maximum likelihood method: SAR, SEM, SLX, SARAR, SDM and SDEM. Their properties were compared in order to choose the one with the best properties. The comparison was based on Akaike criteria, the tests of significance of the parameters λ (measuring the spatial autocorrelation), ρ (measuring the spatial autoregression) and θ (measuring the spatial moving average), likelihood-ratio tests, Wald tests, and Moran and LM tests checking the presence of the spatial processes. Results are presented in Tables 2, 3, 4, 5, 6, and 7.

The construction of spatial weight matrix, used to estimate spatial models, was based on the three-nearest-neighbors method. This method expresses the existence of neighbor relations between observations (NUTS 2 regions). For each region, it ascribes a value of 1 to regions belonging to the set of the three closest neighbors (having a closest centroid) and a value of 0—to other regions. This method was chosen as it works well for the datasets including isolated regions, and the analyzed dataset includes such regions—for instance, Cyprus or Malta.

Table 2 shows that values of coefficients retain similar signs (positive or negative) for the same variables in all compared models. Values of coefficients from SLX, SDM and SDEM models are particularly different as these models include a spatial moving average component, measuring the direct consequence (an impact on excess mortality level) in one region of the cause (a unitary change of one of exogenous variables) in another. Table 3 indicates that these models have also less

Table 2 Coefficients of linear model and spatial models

Variable	Linear	SAR	SEM	SLX	SARAR	SDM	SDEM
Const	−269.90	−233.78	−257.17	−373.40	−231.06	−347.58	−371.44
RIS	−35.63	−38.44	−38.03	−41.17	−34.46	−42.80	−42.90
HumanResources	−116.30	−95.71	−115.41	−114.40	−86.54	−112.85	−111.71
RelationshipOrientation	0.55	0.45	0.53	0.21	0.43	0.20	0.21
LeftHome	45.90	33.85	43.33	31.35	31.89	29.29	29.10
DipIntoSavings	217.40	189.40	216.54	135.50	172.70	136.88	137.14
HappinessJuneJuly	−51.11	−43.80	−51.71	−20.69	−38.89	−21.90	−22.24
LifeSatisJuneJuly	101.10	87.56	100.31	57.13	81.15	57.90	58.67
LonelyChange	7.54	6.52	7.58	5.89	5.93	5.80	5.94
AnxietyRateCOVID	45.84	38.69	44.34	24.58	36.10	24.19	24.74
HealthcareTrustJuneJuly	−15.51	−13.29	−14.93	−8.85	−12.90	−8.86	−9.36
StringencyWeeks27–43	−3.30	−2.76	−3.25	−1.98	−2.54	−1.97	−2.00
PositivityRegion2W	0.62	0.57	0.59	0.59	0.57	0.58	0.58
TestingRegionEnd2W	−0.004	−0.004	−0.004	−0.002	−0.004	−0.003	−0.003
GovTrustChange	20.87	16.69	21.12	11.12	14.10	10.31	10.07
PressFreedomPopul	−2.50	−2.15	−2.35	−0.81	−2.14	−0.76	−0.81
MediaTrustChange	35.25	32.01	34.15	31.32	32.29	31.17	32.54
Percentage65+	263.60	241.64	255.92	198.60	236.98	193.06	194.81

Source own elaboration

Table 3 *P*-values of T-tests of linear model and spatial models

Variable	Linear	SAR	SEM	SLX	SARAR	SDM	SDEM
const	0.0001	0.0003	0.0001	0.0004	0.0001	0.0004	0.0002
RIS	0.0337	0.0145	0.0196	0.0300	0.0215	0.0135	0.0126
HumanResources	2.00E-16	5.40E-12	2.20E-16	1.81E-06	5.53E-09	1.31E-07	1.21E-07
RelationshipOrientation	2.00E-16	2.20E-16	2.20E-16	0.0489	6.33E-13	0.0384	0.0324
LeftHome	0.0078	0.0389	0.0105	0.1465	0.0413	0.1385	0.1382
DipIntoSavings	1.02E-12	8.98E-11	8.55E-14	0.0018	1.02E-08	0.0005	0.0004
HappinessJuneJuly	0.0008	0.0024	0.0007	0.4352	0.0040	0.3681	0.3557
LifeSatisJuneJuly	1.17E-11	6.72E-10	2.22E-12	0.0189	1.22E-08	0.0091	0.0076
LonelyChange	6.89E-13	9.71E-11	5.80E-14	0.0039	6.02E-09	0.0018	0.0012
AnxietyRateCOVID	0.0041	0.0102	0.0051	0.2155	0.0107	0.1836	0.1678
HealthcareTrustJuneJuly	4.48E-06	2.86E-05	9.64E-06	0.2117	1.30E-05	0.1730	0.1449
StringencyWeeks27-43	2.00E-16	1.95E-12	2.20E-16	0.0011	1.79E-09	0.0004	0.0002
PositivityRegion2W	2.19E-08	1.49E-08	1.22E-08	1.63E-06	4.63E-09	1.70E-07	9.76E-08
TestingRegionEnd2W	0.0002	0.0002	0.0002	0.0659	0.0001	0.0427	0.0346
GovTrustChange	0.0019	0.0097	0.0016	0.3243	0.0227	0.3190	0.3247
PressFreedomPopul	0.0001	0.0004	0.0003	0.3857	0.0002	0.3800	0.3387
MediaTrustChange	0.0028	0.0037	0.0035	0.0857	0.0015	0.0619	0.0479
Percentage65+	3.83E-08	3.84E-08	2.24E-08	0.0007	1.91E-08	0.0003	0.0002

Source own elaboration

Source own elaboration

Table 4 Comparison of the properties of one-source spatial models

Test	SAR	SEM	SLX
p-value (Moran test on the presence of spatial process)	0.748	0.467	0.8004
p-value (likelihood-ratio test)	0.007	0.239	
p-value (Wald test)	0.005	0.189	
Akaike criterion	−3296.9	−3291.0	

Source own elaboration

Source own elaboration

significant variables at the 5% significance level—depending on the model the number of statistically significant variables is of 10 and 11, while all other models have 18 significant variables, similarly to a linear model.

Source own elaboration

Tables 4 and 5 contain information about significance of ρ and λ parameters, residuals autocorrelation and the quality of models compared to the linear model. A spatial regression parameter ρ is significant only for the SARAR model, and the result of for the SDM model indicates that SLX has an advantage over it. At the

Table 5 Comparison of the properties of two-source spatial models

Test	SARAR	SDM	SDEM
p-value (test concerning ρ)	0.005	0.287	
p-value (Wald test for ρ)		0.271	
p-value (test concerning λ)	0.200		0.374
p-value (Wald test for λ)			0.329
p-value (Moran test)	0.469	0.529	0.461
p-value (likelihood-ratio test)	0.013		
Akaike criterion	−3296.4	−3284.7	−3284.3

Source own elaboration

Table 6 LM—testing the presence of the one-source spatial process

Test	H0	Restriction	H1	p-value	Decision
Lmerr	Linear	$\lambda = 0$	SEM	0.276	Linear
Lmlag	Linear	$\rho = 0$	SAR	0.008	SAR
RLMerr	Linear	$\lambda = 0$	SEM	0.198	Linear
RLMlag	Linear	$\rho = 0$	SAR	0.006	SAR
SARMA	Linear	$\lambda = 0$	SARAR	0.013	SARAR
		$\rho = 0$			

Source own elaboration

significance level of 0.1, parameters λ were significant for none of the two-source models indicating an advantage of SAR model over SARAR and an advantage of SLX over SDEM. Moran tests for residuals indicated that there is no presence of spatial processes for any of the models. Nevertheless, likelihood-ratio tests show the better fit of SAR and SARAR models compared to the linear model. Wald test for SAR model indicated also the significance of the ρ parameter. In the case of the SEM model, this test showed the insignificance of the λ parameter. According to Akaike criteria, model SAR and SARAR are of the best quality. These two models have also lower Akaike criteria than the linear model—they amount to respectively −3296.9 and −3296.4 against −3291.6 of linear model.

Tables 6 and 7 enhance the conclusion of the best fit of SAR and SARAR models and their advantage over the linear model, which is linked to the significance of ρ and λ components. The first likelihood-ratio test showed that the SAR model reflects reality in a better way than SARAR—which strengthens the conclusion related to Akaike criteria values.

The SAR model seems to be the best also from a theoretical point of view, as it analyzes a relationship between independent variables and the dependent variable at the consequence level. Thus, it analyzes the direct connection between mortality

Table 7 Likelihood-ratio tests—testing the presence of one-source and two-source spatial processes

H0	H1	p-value	Decision
SAR1	SARAR	0.222	SAR
SEM1	SARAR	0.007	SARAR
SAR1	SDM	1.89E-05	SDM
SEM1	SDM	9.96E-07	SDM
SLX1	SDM	0.287	SLX
SEM1	SDEM	−1.18E-06	SDEM
SLX1	SDEM	0.374	SLX

Source own elaboration

levels of different regions. Considering the specificity of the endogenous variable, this approach should be more consistent with reality than the linear model in which the connection is analyzed at the level of causes (exogenous variables). Similarly, the SLX model (estimating a direct impact of a cause in one region on a consequence in another) will be better for the model having excess mortality rate as a predictor and not endogenous variable. Coefficients and *p*-values of related T-tests of the selected SAR model are presented below.

Spatial stimuli calculated on the basis of Model 4 were used to illustrate impacts of the change in regional innovation index in one region on the excess mortality in all other regions. Visualizations prepared for six selected Italian, French and Swiss regions are presented in Figs. 7, 8 and 9.

Figure 7 shows the impact of the unitary change in regional innovation index value in Lombardy, Liguria and the Aosta Valley on excess mortality (from weeks 38 to 47 of 2020) in other analyzed regions. Regions with a greater impact (marked in green) are scattered, which may be related to the fact that the model does not take into account certain factors such as tourism attractiveness or the balance of migration. As can be seen, meaningful relations may be difficult to capture as the SAR model is based on the relationships between mortality levels in different regions. It indicates, however, closer links between certain capitals and other major cities, such as Milano, Torino, Florence, Madrid, Barcelona, Paris, Stockholm, Prague, Budapest and Bucharest. Moreover, it shows the possibility of existence of stronger relations between Northwestern Italy and such European capitals as Madrid and Paris. Cities of Milano, Madrid and Paris are also the biggest luxury retail cities of the European Union based on share of global luxury store openings and total retail sales in the highly-innovative luxury sector.[67]

Figures 8 and 9 show that in the case of some regions, impacts of the change in their innovation level value on excess mortality is concentrated on other regions from the same country or from a common linguistic area. In the case of Swiss Lemanic Region (or Lake Geneva region), the impact on French Champagne, and Belgian Namur and Hainaut regions may be partially associated with the wide wine production in all of these regions—Lemanic Region is responsible for 68% of the

---

[67]Savills Global Luxury Retail 2019 Outlook [57].

Model 4 The SAR model explaining excess death number per 100,000 inhabitants in 2020 (weeks 38–47)

Variable	Coefficient	p-value
Const	−233.780	0.0003
RIS	−38.441	0.0145
HumanResources	−95.709	5.40E-12
RelationshipOrientation	0.453	2.20E-16
LeftHome	33.853	0.0389
DipIntoSavings	189.400	8.98E-11
HappinessJuneJuly	−43.795	0.0024
LifeSatisJuneJuly	87.557	6.72E-10
LonelyChange	6.520	9.71E-11
AnxietyRateCOVID	38.690	0.0102
HealthcareTrustJuneJuly	−13.292	2.86E-05
StringencyWeeks27–43	−2.756	1.95E-12
PositivityRegion2W	0.567	1.49E-08
TestingRegionEnd2W	−0.004	0.0002
GovTrustChange	16.686	0.0097
MediaTrustChange	32.011	0.0037
PressFreedom	−2.152	0.0004
Percentage65+	241.640	3.84E-08

Source own elaboration

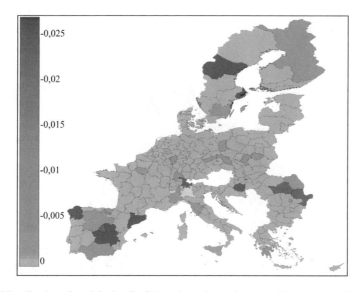

Fig. 7 Visualization of spatial stimuli of the unitary change in regional innovation index value in Lombardy, Liguria and the Aosta Valley on excess mortality (from weeks 38 to 47 of 2020) in other analyzed regions. *Source* own elaboration

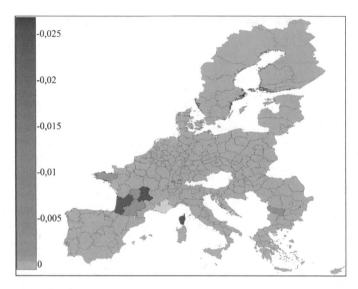

Fig. 8 Visualization of spatial stimuli of the unitary change in regional innovation index value in Provence-Alpes-Côte d'Azur on excess mortality (from weeks 38 to 47 of 2020) in other analyzed regions. *Source* own elaboration

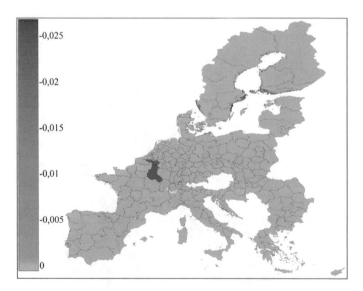

Fig. 9 Visualization of spatial stimuli of the unitary change in regional innovation index value in Lemanic Region on excess mortality (from weeks 38 to 47 of 2020) in other analyzed regions. *Source* own elaboration

national wine production in Switzerland,[68] Champagne is the most famous and one of the main wine producing regions of France and Namur and Hainaut encompass the largest Belgian wine producing region, Côtes de Sambre et Meuse.

A limitation of spatial stimuli matrix to impacts directed to regions from other countries than the country of stimulus origin allows for additional conclusions. Before juxtaposing regions with the highest values of international impacts, 47 regions were excluded from the analysis. These regions were previously artificially aggregated to form a bigger macro-regions in the process of the construction of variables *LeftHome*, *DipIntoSavings* and *AnxietyRateCOVID* to ensure reliability of the results and to overcome the problem of a small number of respondents from certain regions. Now, these regions could be responsible for distorting conclusions related to spatial stimuli. The remaining 188 regions were subsequently sorted in a descending order starting from the region with the biggest impact of innovation index change on excess mortality in other regions from different countries than the country of impact origin.

The first twelve regions included Bucharest-Ilfov region, Brussels, Malta, Groningen province, Eastern Switzerland, Eastern Slovenia, Champagne, East Flanders, Central Switzerland, Latvia, Espace Mittelland and Ticino. First conclusion is that an analysis on the NUTS 3 level could bring more valuable results as the comparison of NUTS 2 regions include both very small regions consisting of one major city, such as Bucharest and Brussels that have a greater chance to impact surrounding areas, and larger regions or even whole countries. Secondly, regions comprising more important cities are more likely to have a higher place in the classification, as 8 out of the remaining 24 country capitals are in the first 15 regions of the ranking, but some of them belong to the regions ranked in the bottom part of the classification. In particular, this is the case of Warsaw, Berlin and Athens, situated in regions which are among 50 NUTS 2 with the lowest international impacts on excess mortality. Furthermore, different factors which were not included in the model, such as migration and touristic attraction, can be relevant, considering a high rank of Malta.

The next conclusion is related to the high performance in this ranking of Benelux and Swiss regions, scoring the highest in Europe in terms of regional innovation. Four out of seven Swiss NUTS 2 regions are located in the top 12 places of the international-impact ranking, other regions being also near the top. Capitals of other top-scoring provinces include Groningen (having the largest number of successful start-ups in country after Amsterdam,[69] and with the highest level of satisfaction with educational facilities and healthcare system in Europe[70]), Ljubljana (2016 green capital of Europe and the former first European capital in terms of the level of satisfaction with educational establishments[71]), Gent (having second best ranked

---

[68]Grevet [31].

[69]A look at the startup boom in Groningen [1].

[70]Quality of Life in European Cities [54].

[71]Ibid.

university in the country), Riga, Porto, Vilnius and Rome. Educational facilities are also of a high performance in Sankt Gallen, the main city of the first Swiss region—Eastern Switzerland, as its university is ranked first in the country. Other high-scoring regions in the international-impact classification are German Bayern and Freiburg, also performing well in the field of education occupying 1st and 15th place among the EU cities with highest ranked universities of 2020.[72] The second place in Italy and 29th place among all regions in the international-impact classification belongs to Milano, which is among two the most important EU cities for a highly-innovative luxury sector. This indicates that, among other factors, knowledge hubs, propitious for a creation of the start-up culture, can be important in interregional attempts to mitigate the effects of such challenges as COVID-19 pandemic.

7 Silver Lining of the Pandemic that Reshapes the Luxury Industry

Considering the aforementioned results, patrons should look closer at the luxury industry and its role in shaping the post-COVID world. It is believed that this is one of the most, if not the most, affected industries in times of crisis. Briefly reviewing the past crises, explains how this is true for the industry. Comparisons to previous crises show similarities in experiences from then, to the latest crisis mankind have ever faced—the COVID-19 pandemic. This section shall speak about -SARS (SARS-CoV-1), MERS and SARS-CoV-2 (causing COVID-19), which were all significant, as the year of its conception, has taken a toll on different aspects of life, including human health, social and economic.

Severe Acute Respiratory Syndrome (SARS) lasted from the first quarter to the third quarter of 2003. Similarly to COVID-19, this health crisis originated also in China with a high impact on global economies. Reported to have emerged as early as 2002, SARS epidemic slew China's economy down by 2 percentage points compared to the same period of Q1 2002 and 2003, with 11.1 and 9.1% of economic growth respectively.[73] Retail was one of the hardest hit sectors, with a 4.3% decrease in sales reported in May 2003. Already during this time, non-essential goods, classified between luxury and durable goods like clothing, were more affected than other categories such as food and beverage. Yet, framing the context here, where the virus did not spread in a greater geographical scale and where China was not yet an economic powerhouse as it is today, luxury continued to thrive in major economies like Japan and the United States. However, the table below (refer to Fig. 1) presents how quickly the industry was able to recover in the case of China. The apparel category was quick to rebound just two months after the

---

[72]Best universities in Europe [7].

[73]Moy [46].

infection rates started to drop mainly due to repressed consumer spending which was already starting to heighten in the region.

The SARS pandemic had a strong impact on retail as it influenced decision makers and retailers in general to embrace e-commerce as an alternative growth channel.

Another outbreak was experienced shortly after with a very small geographical spread—Middle Eastern Respiratory Syndrome (MERS). Breaking out in South Korea in 2015, MERS pandemic lasted only for less than a year, with major economies impacted and again the retail sector took one of the biggest hits. The quick recovery, however, is clearly visible in the table below. The illustration mentions the Mid-Autumn Festival that coincided with the outbreak spread easing down (Figs. 10 and 11).

In 2007–2009, a financial crisis caused a big wave of economic consequences around the globe. The Great Recession was probably the closest scenario that can be compared to the COVID-19 pandemic, however, depending on the country and their level of debts, the impact of this recession was nothing short of massive. This crisis may not have caused store closures, as that of COVID-19, however customer spending behaviors were felt. Tracing back to this period, e-commerce strategies were not solidified yet, forcing marketers and consumers to reevaluate their strategies and their lives respectively. Working hours, minimum wage and unemployment rates are some indicators that were results of this global downturn.[74]

In summary, these three waves of crisis in forms of health and finance, and the effects of these scenarios have continuously shaped ways of doing business across industries. Since the SARS crisis has started in Asia and the effects of it were first felt hard in the region, more specifically in Asia's biggest market, China, it may explain why this region has adapted better compared to other parts of the world. By better, it is defined as establishing patterns for faster control and recovery both by the government and the private sectors across all related countries. In the same way, it has undoubtedly better prepared for future crises such as the one the world is facing today. Most probably, the same goes for MERS in South Korea, that truly helped the country to adapt to COVID-19 today.

The success stories on recovery and growth witnessed during the outbreak from these two health related pandemics, seem to follow pursuits in countries with strong economies that are likely to see a surge in retail spending once the virus spread is controlled. In high-developed economies, both pandemics have also led to quicker actions taking both the public and private offices and their respective businesses in context.

Regarding the ongoing coronavirus pandemic, the way people live and work has definitely been transformed to almost unimaginable ways at the beginning of 2020. It may seem counterintuitive to even discuss the future of luxury and discretionary spending in a time of financial crisis as the unemployment rate soars and even the world's most powerful markets struggle to stay relevant and/or afloat. But amid the

---

[74]Cf. Verick and Islam [64].

The Impact of SARS (1) on Retail Markets

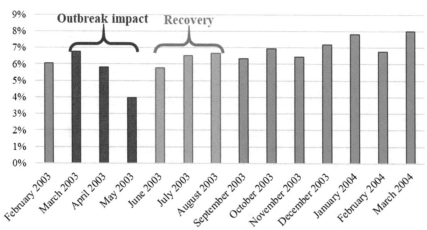

Fig. 10 The impact of SARS on retail markets. *Source* own elaboration based the data of Bain & Company and National Bureau of Statistics of China. Focus on China, where the outbreak took place from March to May 2003, steady recovery was noted shortly after the easing down of the spread

The Impact of MERS on Retail Markets

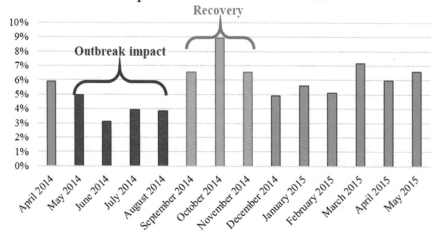

Fig. 11 The impact of MERS on retail markets. *Source* own elaboration based the data of Bain & Company and National Bureau of Statistics of China. Focus on South Korea, where the outbreak was realized from May to August 2015, rapid recovery was notable also because of the coincidence of the celebration of a local festival

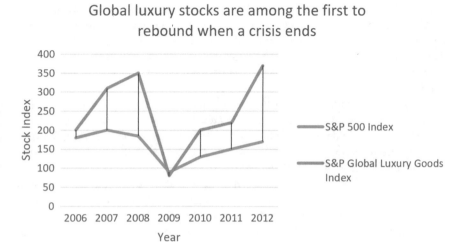

Fig. 12 Global Luxury stocks are among the first to rebound when a crisis ends. *Source* S&P Global, Luxe Digital. Consequent effects of outbreaks or disruptions in the economic cycles are felt in the stock market, highlight the aftermath for the global luxury industry from 2006 to 2012

high level of uncertainties, the pandemic has been a turning point for all industries and lifestyles to change and adapt to more peculiar, aware and careful consumer behaviors. This altogether, creates new imperatives for all of the luxury sectors (Fig. 12).

While the retail sector has always been the first to take the hit, it has also been the first one to rebound from the crises. Be it a health or financial crisis, the luxury industry was quick to respond and learn on how to become more customer-centric in their approach to better engage consumers while also putting a lot of effort into becoming more digital in their ways. These two concepts, used as strategies, catalyze innovation to be at the industry's core. And because of its innate attributes in creativity, used mainly to mold and promote its business, it is as if the industry has its fair hand to lead the discussion on innovation and sustainability. And in theory, be responsible to take actions on making one complement the other.

At the beginning of the pandemic, multinational fashion and luxury houses converted their manufacturing plants to respond to consumer's needs. Such responses did not involve the usual luxury goods but the essentials. International luxury groups—Richmond Group and Ralph Lauren offered financial assistance to both hospitals and non-profit organizations. Meanwhile, Bulgari, Armani Group, LVMH and L'Oréal decided to convert their production sites to assist in producing essentials such as hand sanitizers, hospital gowns and face masks. In the face of a global pandemic, these large groups and well-known brands, were quick to show solidarity with their patrons and communities, pivoting to address urgent health needs. With such decisions, loyal followers find relevance to the brands they patronize. The decision is hinged on creatively finding relevance to create brand awareness. However, beyond immediate reactive initiatives, luxury brands will continue to further adapt their answer to their customers' most pressing needs by revamping their product mix and service delivery in alignment with today's new reality.

In light of the current state of the world, motivations demonstrated by the new consumers are now non-negotiable and the industry is forced to respond. Brands are grappling with a fundamental shift in what "luxury" means, as consumers become more environmentally and socially aware, and digital channels become more important as sources of inspiration and sales.

In fact, according to *Sustainability and Cooperation to Dictate the Future of Companies*, in 2019 alone, 80% of Millennials and Gen-Z stated that their purchase decisions would be influenced by sustainability. Needless to elaborate on this matter, this is true and felt by the industry. Most notably, due to the number of luxury consumers who belong to this generation cluster and their consciousness to the matter, such components in businesses are necessary to thrive in the new paradigm. Companies that respond by streamlining operations, redefining luxury to be less conspicuous and more inclusive, and investing in new ways of doing businesses are more likely to not just gain power through these uncertain, changing times, but also emerge stronger for the future.

At this point, it will be discussed what were the trends and developments that luxury groups and brands have continued reinventing themselves and how they are responding through the call of the new generation of luxury goods consumers. Some topics that will be highlighted in this section are not just present, applicable and actionable steps to move forward, but also what is thought to be missing dialogues whose attention has been long overdue. All thanks to the COVID-19 pandemic, developments for these sustainability related actions were accelerated.

8   Purpose and Intent as Luxury's New Legacy

Luxury brands are inherently of the highest quality. This excellence in craftsmanship is earned and should consistently be delivered. Beyond a high-quality product, though, luxury brands will have to not only display strong values, but also deliver purposeful actions and build a sense of close-knit community to gain the trust of discerning, conscious consumers, and forge deeper connections.

Throughout time, man has continuously discovered and invented ways on how it can better live and conveniently support lifestyles. It will not be easy to disrupt what humans have already been accustomed to. Businesses that are already running, profits that are continuously being analyzed and growing, and traditions that are kept celebrated. Specifically, the jewelry sector more precisely the diamond industry has a lot of valuable inputs to share here. Dating from 1999, when an advertising campaign was launched by the De Beers, "A Diamond is Forever", society was made to believe that every marriage must be complemented by a diamond ring. The brilliant minds behind this campaign were not just riding on an already existing proposition, but rather, started to fuel this idea.[75] Until today, society continues to celebrate a tradition of marriage complemented by a proposal ring different from a marriage ring. It continues to evolve, as even same-sex marriages now across different societies are accepted, while friendship or promise rings are becoming a trend, and all that consequently seeks for a higher demand on the availability of these precious gems, diamonds.

There is now the development of lab-grown diamonds, serving purpose as its new product. Zooming onto developments in the diamond-making industry, lab-grown diamond is slowly taking a piece of the pie in the industry of jewelry. Processed diamonds, that were experimented 70 years ago and were then continuously explored, have received a certification from the Gemological Institute of America (GIA), proving that the components not limited to the purity, density and color, are comparable if not better than the natural made gems. This milestone on the discovery of lab-grown diamonds, luxury consumers and enthusiasts turn heads to, may soon be a more sustainable and attractive alternative for a more sustainable purchase.[76]

The lab-grown diamonds' environmental impact is a feat towards 'Conscious Luxury'. When it comes to the water usage, mined diamonds absorb an estimated 126 gallons of water per carat, compared to the lab-grown diamond whose water absorption is significantly lower—at only 18 gallons. One can imagine where does all the water waste go for mine diamonds, and according to *Grown Diamonds Key to Unlocking Future for Diamond Industry*, all these wastewater and pollutants end up at the surface of water bodies.[77] Similarly, concerning the energy usage, natural diamonds at its most current and traditional way of generating, would require resources that include diesel fuel, electricity and hydrocarbons that all release harmful carbon into the air, while its alternative, lab-grown diamonds could work on renewable energy to operate machines that directly develop these fine jewels. With this information at hand and few others available on land disruption and waste generation, luxury finds its way to a more progressive approach to 'Conscious Luxury'.

Finally, the carbon emission of a single carat will emit 125 lb of carbon for a naturally mined diamond while only 6 lb of carbon for the lab-created one. Luxury consumers continue to become more environmentally and socially aware—thus, becoming more diverse, more inclusive, and more sustainable and ethical will be non-negotiable.

---

[75]Kolowich Cox [35].

[76]Lab Grown Vs Mined [37].

[77]Brown [9].

From the fashion side as well, it is evident that there are a lot of efforts done to attain sustainability in the existing operations model. A trend that is really attractive is resale of secondhand purchased goods. This includes products, not limited to apparel, taken from the thrift and donation channels. Inspired by the problem which exists but is easily neglected in fashion waste, several platforms thrive on other people's trash. Fashion waste crisis exists. In fact, according to *What retailers need to know about resale in fashion*,[78] the industry ranks number 3 as most polluting in the world, just behind fuel and agriculture. Fashion production makes up to 10% of humanity's total carbon emissions. The report explains further that the attributes considered here are coming mainly from 'fast fashion'. This terminology refers to clothing or garments made from cheaper material, most often then—not containing microplastics. Particular brands are famous for the 'runway to rack' concept where faster stock turn-around while selling higher volumes at inexpensive price points constitute the main business proposition. It is worth noting that manufacturing plants of concerned brands and business models are deployed in Asia where factories run mostly on coal and gas, which altogether affects the total CO_2 computation. Citing another research entitled *What Retailers Need to Know About Resale in Fashion*, in the one of the most sought-after online marketplaces in the resale sector, 75 lbs. of CO_2 is produced for an average pair of jeans, 700 gallons of water are required to make a new T-shirt, and 1 in 2 people are disposing unwanted clothes straight to the garbage bins. The result of the latter explains why approximately 64% of the 32 billion garments produced each year end up in landfills. At the same time, combustion of coal or other fossil fuels releases toxic elements and particles with a diameter of 2.5 and 10 μm.[79] In Europe, carbon consumption per capita is positively correlated with the PM10 measure explained earlier.[80]

As this existing concern prevails, the cost of a single-use item to the environment, in theory, contributes to the 208 million lbs. of waste generated out of the single-use outfits in 2019. The table presenting PM10 emissions does not capture the fashion industry alone, but from the figures, it can be derived that approximately 10% of particulate matter comes from the fashion industry. While this problem is highlighted, creativity (of the industry or the enthusiasts who run and develop it) was at play here. Innovating the way to approach a problem while running a business was the core of the following proposed solutions. Secondhand vintage, rental clothes or simply-reusing clothes are examples of developing ideas around this crisis in fashion.

"As the momentum to solve the climate crisis builds, consumers are waking up to the realities of fashion's impact on the environment. Shopping second hand is one of the most effective ways for us to collectively lower our fashion footprint and make the most of the resources used to create these garments."—Elizabeth L. Cline, author of The Conscious Closet.

---

[78]Trotter [61].

[79]Pył zawieszony: czym jest PM10, a czym PM2.5? Aerozole atmosferyczne [52].

[80]Own elaboration based on the data of World Health Organization and Eurostat.

Furthermore, resale, understood as both a concept and a thriving business model, is at play. Given the demographics in the developing online space, this is a hype for the new and more conscious generation, the Gen-Z. Thus, resale is expected to be key in the future of fashion. The sector is expected to accelerate in the next 5 years while retail continuously diminishes. In 2019 alone, resale grew 25 times faster than the traditional retail sector, while retail clothing decreased by 4 percentage points, according to the *GlobalData Market Sizing and Growth Estimates from 2019 to 2024 Projected Market Growth* study. All these concepts combined, resale can be expected to take over fast fashion by 2029. ThredUp reported to have recirculated 4 million fast fashion items—equivalently 5.5 million lbs. of waste.[81] As the market either demands progressive change towards the subject or the maturity to accept secondhand items continuous growth, more brands are now responding to this call by opening their business models to the idea of this new trend (Fig. 13).

Luxury has been flirting with sustainability for years now. However, growing environmental, social and governance concerns will put the luxury industry's sustainable efforts under heavier scrutiny—reinforcing the need for luxury brands to become more trustworthy by providing more transparent information about their processes and products. Even in Asia, where the majority of the luxury consumers are hailing from, all thanks to the countries' rising middle class, status symbol significance and social capital, which are still traditional motivations for purchasing luxury for many first luxury shoppers. It will be interesting to see how luxury fashion houses will try to influence and adjust to this go-to market where mindset is yet to fully adapt to sustainability.

In a movement towards sustainability, Burberry's partnership with the luxury online reseller, The RealReal, has become an example of the growing acceptance of the secondhand clothing market. The InsiderTrends reports that for a specific period, customers reselling Burberry products were given invitations to high tea and a personal shopping experience in the Burberry store itself. This partnership has proven to be beneficial to the luxury brand as The RealReal reported a 64% year-on-year increase in resale interest for Burberry. Interestingly, Stella McCarntey has also ventured into the resale trend with The RealReal to help extend the life of its products. Customers who consign their Stella McCartney products with The RealReal receive store credit to use with the brand on new purchases.

It seems that these trends or developments are slowly and more fundamentally present to modify the existing luxury landscape to become a 'Conscious Luxury' model. The COVID-19 pandemic did a lot of disruption in the reevaluation of what the industry has delayed for so long. The pandemic sped up this shift and acted as a catalyst of change. From Europe to Asia, countries amongst regions across the world experience these changes differently in terms of speed and context. It should be highlighted again that Asia as a region has had similar health crises in the past that allowed the leaders of the regions to plan more proactively to keep economies afloat while being agile to minimize the destructive impact of crises. Meanwhile,

---

[81]Fashion Resale Market and Trend Report [29].

Fig. 13 New brands partner with a resale online marketplace. *Source* own elaboration based on ThredUp graphic. ThredUp illustrates how more brands throughout the timeline joins the trend on resale, clearing showing the maturity of the market both the consumer and the seller (brands) to reuse and resell previous purchases and/or more vintage goods

Europe and the US levels of uncertainty remain high, and could potentially remain as such well into the year 2021.[82]

Current mindsets and challenges will be tackled to further explore creativity and innovation made in the industry, that highlights relevancy as the industry's new legacy. Simply, finding relevance at current times but while finding balance on essentialism and heritage.

There is support coming from investors turning their attention to renewable energy and companies finding ways to partner with right suppliers and vendors that can reach quickly and at scale. Luxury groups will have the lee way to keep redesigning processes to shorten lead time and sharpen innovation while keeping quality at high levels. Investors are now considering Environment, Social and Governance (or ESG) Investing—as both profitable and beneficial to society. Assets related to such investing are forecasted to grow from 11% only in 2012, up to 50% in 2025. In December 2019, Goldman Sachs announced a 750-billion-dollar plan over the next decade to finance and advise companies focused on sustainable finance themes. More recently, in January 2020, BlacRock released a letter concerning sustainability addressed to all their clients, sharing their vision of doubling the number of ESG exchange traded fund offerings and making ESG analysis a key component of portfolio construction. These are developments and conversations already happening, and the industry has a greater playground to take advantage of

---

[82]Zipser and Achille [68].

them, given that all the concerns and possibilities concerning both waste as a problem and innovation as a solution are present in the current context.

Data analysis will play a larger role in such intelligence gathering, allowing brands to create and deliver the right merchandise to the right place at the right time. It is also expected that the industry will undergo a strong consolidation, with size and scale becoming even more important than they were in the previous decade. Despite all the uncertainty of this new normal, brands that spot opportunities, where others only see challenges, will not only survive but thrive in the new normal. Because, in the end, the only insurance for luxury brands (and all brands for that matter) is to remain relevant and in alignment with their consumers' evolving expectations. Independent brands whose efforts are focused at the local level from sourcing, to production and finally to distribution are the fitting examples to this. Jonas Hoffman, in his book *Independent Luxury*, cited the trend of operating locally while thinking globally. Which in its very essence, is how the industry started prior globalization.

Brands in the industry will have to reflect on their inner essence to provide authentic pieces that are culturally accepted and socially relevant. Brands will have to find balance between their roots from the past concerning rarity and craftsmanship and the projection of the present revolving around consciousness and sustainability. Luxury, as it gravitates towards heritage, will remain creative to seek for other avenues to find relevancy. Beyond the heritage and their historical credentials, the luxury industry will have to adapt itself to today's expectations.

In June 2020, Gucci released its first experiment in circularity, Gucci Off the Grid, a line of accessories and streetwear made from recycled, organic, bio-based and sustainably sourced materials. The younger generation, which will make up the majority of the Luxury consumer base, will seek for more sustainable sourcing from the global brands. According to Bain and Company's *Eight Themes That Are Rewriting the Future of Luxury Goods*, 80% of luxury customers say they prefer brands that are socially responsible particularly among millennials. Furthermore, 60% of luxury customers think that luxury brands should be more engaged in sustainability than other industries.

9   Conclusion

The key findings of this chapter revolve around both current and future role of innovation in the COVID-19 pandemic and post-COVID world. An econometric analysis showed that a relatively high innovation level, including the quality of human resources (measuring the level of civil society) and intellectual assets, may mitigate the effects of the pandemic. An excess mortality rate is also likely to be influenced by the culture specificity, especially relationship orientation and long-term orientation. Other determinants of anti-COVID measures effectiveness are related to the behavior changes such as staying at home and psychological aspects—a tradeoff between loneliness severity and policy effectiveness was identified. Another finding is related to the higher anxiety and nervousness associated with the increased excess mortality anomaly. Furthermore, a deterioration of economic situation of modest households associated with the 2020 crisis makes it more difficult to deal effectively with the pandemic effects. On the other hand, a more rigid lockdown stringency and a greater testing rate go in line with a lower mortality anomaly in Europe. Not surprisingly, the importance of the quality of governing and healthcare system should not be neglected. One can also pay a particular attention to the lower efficiency of populist authorities. The econometric analysis indicated that, in the context of coping with the pandemic, a more harmful feature of populist governments is related not so much with the restrained media freedom itself (one of key characteristics of such governments), but rather with the poor media quality (reflected by the changing level of trust towards media). Higher excess mortality rate in 2020 was also connected with the increased percentage of people aged 65 years and more. Interestingly, a relatively high death anomaly in the first wave of the COVID-19 pandemic turned out to be, *ceteris paribus*, related to the higher excess mortality during the second wave. Another conclusion illustrates that a poor air quality may extend the duration of the pandemic wave after its peak is reached—particulate air pollution is likely to prolong the fading phase of the pandemic or make it more severe. One of the most important findings provided by the SAR econometric model indicated that, among other factors, knowledge hubs (propitious for a creation of the start-up culture) can be important in interregional attempts to mitigate the effects of such challenges as COVID-19 pandemic.

With regard to the undisputed importance of innovation in the post-COVID world, the chapter presented as well the role of the luxury sector. Luxury, at its core, offers superior value and has an inherent ability to make timeless emotional connections in people's minds. Many affluent shoppers will be reverting to less conspicuous, quieter forms of luxury. In these unprecedented turbulent times, more intimate values of luxury may prove to be more relevant than ever. And the change is sure to come. The very meaning of luxury is also set to become much more diversified and contextual. The year ahead will call on luxury brands to evolve and adapt in accordance with the changing societal landscape. Luxury brands will need to revisit basic assumptions and the traditional luxury playbook, learn new vocabularies, and innovate. The concept of revisiting its roots by the luxury industry concretizes the ideas behind the reality of our society's current state and the expectations thereafter due to global climate change. Such a decision to transition to greener solutions entail time and resource and yet yields opportunities. The luxury industry moves towards this direction and, although the transition was rather forced given the current context (which may be the case of any other industry), yet the right mindsets, tools and investments are accessible to keep related developments efficient and effective in the long run. Luxury as a culture and as a strong business will continue to evolve and, as it does, its storytelling should be, therefore, about celebrating consumer's passions and enabling self-expression, not simply telling a brand history.

References

1. A look at the startup boom in Groningen (2016) [Online]. StartupJuncture. Available at: https://startupjuncture.com/2016/03/18/a-look-at-the-startup-boom-in-groningen/. Accessed 12 Feb 2021
2. Adler P, Seok-Woo K (2002) Social capital: prospects for a new concept. Acad Manage Rev 27(1)
3. Allen J (2020) Fauci says herd immunity could require nearly 90% to get coronavirus vaccine [Online]. Reuters. Available at: https://www.reuters.com/article/health-coronavirus-usa/fauci-says-herd-immunity-could-require-nearly-90-to-get-coronavirus-vaccine-idUSL1N2J411V. Accessed 12 Feb 2021
4. Bahmani A (2020) Early detection of COVID-19 at scale using wearables [Online]. Nature Research Bioengineering Community. Available at: https://bioengineeringcommunity.nature.com/posts/early-detection-of-covid-19-at-scale-using-wearables. Accessed 12 Feb 2021
5. Ball P (2021) The lightning-fast quest for COVID vaccines—and what it means for other diseases, vol. 589. Springer Nature
6. Basith S, Manavalan B, Shin T, Lee G (2020) Machine intelligence in peptide therapeutics: a next-generation tool for rapid disease screening. Med Res Rev 40(4)
7. Best universities in Europe (2021) [Online]. Student. Available at: https://www.timeshighereducation.com/student/best-universities/best-universities-europe. Accessed 12 Feb 2021
8. Blondeel M (2020) COVID-19 and the climate—energy nexus. Eur Policy Brief (61)
9. Brown A (2016) Grown diamonds key to unlocking future for diamond industry, finds frost & Sullivan [Online]. Frost & Sullivan. Available at: https://ww2.frost.com/news/press-releases/grown-diamonds-key-unlocking-future-diamond-industry-finds-frost-sullivan/. Accessed 30 Jan 2021
10. Burke CW (2020) Using AI to find peptide therapeutics for COVID-19 [Online]. BioSpace. Available at: https://www.biospace.com/article/using-ai-to-find-peptide-therapeutics-for-covid-19/. Accessed 12 Feb 2021
11. Chandler JD, Graham JL (2010) Relationship-oriented cultures, corruption, and international marketing success. J Bus Ethics 92(2)
12. Clausing K (2020) The case for economic openness in the time of coronavirus [Online]. The CGO—Center for Growth and Opportunity at Utah State University. Available at: https://www.thecgo.org/wp-content/uploads/2020/09/The-Case-for-Economic-Openness-in-the-Time-of-Coronavirus.pdf
13. Compendium of OECD Well-being Indicators (2011) [Online]. OECD. Available at: http://www.oecd.org/sdd/47917288.pdf
14. Country Comparison [Online]. Hofstede Insights. Available at: https://www.hofstede-insights.com/country-comparison/. Accessed 12 Feb 2021
15. Country navigator—your tool for global skills development (2018) [Online]. The Culture Mastery. Available at: https://theculturemastery.com/cultural-and-personality-assessment-tools/country-navigator/. Accessed 12 Feb 2021
16. COVID-19 government response tracker (2020) [Online]. The Blavatnik School of Government. Available at: https://www.bsg.ox.ac.uk/research/research-projects/coronavirus-government-response-tracker
17. COVID-19 recovery in hardest-hit sectors could take more than 5 years (2021) [Online]. McKinsey & Company. Available at: https://www.mckinsey.com/featured-insights/coronavirus-leading-through-the-crisis/charting-the-path-to-the-next-normal/covid-19-recovery-in-hardest-hit-sectors-could-take-more-than-5-years. Accessed 12 Feb 2021
18. COVID-19: stringency index (2020) [Online]. Our World in Data. Available at: https://ourworldindata.org/grapher/covid-stringency-index?stackMode=absolute&time=2020-08-30&region=Europe

19. Dakhli M, De Clercq D (2004) Human capital, social capital, and innovation: a multi-country study. Entrepreneurship Reg Dev 16(2)
20. D'Arpizio C, Levato F (2020) How Covid-19 has accelerated the transformation in luxury [Online]. Bain & Company. Available at: https://www.bain.com/insights/how-covid-19-has-accelerated-the-transformation-in-luxury-video/. Accessed 12 Feb 2021
21. Data for the maps in support of the council recommendation on a coordinated approach to the restriction of free movement in response to the COVID-19 pandemic in the EU/EEA. (2021) [Online]. European Centre for Disease Prevention and Control. Available at: https://www.ecdc.europa.eu/en/publications-data/indicators-maps-support-council-recommendation. Accessed 12 Feb 2021
22. Data on testing for COVID-19 by week and country [Online]. European Centre for Disease Prevention and Control. Available at: https://www.ecdc.europa.eu/en/publications-data/covid-19-testing. Accessed 12 Feb 2021
23. Data on the weekly subnational 14-day notification rate of new COVID-19 cases. (2021) [Online]. European Centre for Disease Prevention and Control. Available at: https://www.ecdc.europa.eu/en/publications-data/weekly-subnational-14-day-notification-rate-covid-19. Accessed 12 Feb 2021
24. Deaths by week, sex and NUTS 2 region (2021) [Online]. Eurostat. Available at: https://appsso.eurostat.ec.europa.eu/nui/show.do?dataset=demo_r_mwk2_ts&lang=en
25. DGP: RFIL rozdzielony według klucza politycznego (2021) [Online]. Serwis Samorządowy PAP. Available at: https://samorzad.pap.pl/kategoria/aktualnosci/dgp-rfil-rozdzielony-wedlug-klucza-politycznego. Accessed 12 Feb 2021
26. Directorate-General for Economic and Financial Affairs of the European Commission (2020) European economic forecast—autumn 2020. European economy institutional papers, 136
27. EU trade visualisation (2021) [Online]. PBL Netherlands Environment Assessment Agency. Available at: https://themasites.pbl.nl/eu-trade/index2.html?vis=chord. Accessed 12 Feb 2021
28. Exposed: Slovenian government exploits pandemic to shatter environmental safeguards (2020) [Online]. BirdLife Europe and Central Asia. Available at: https://www.birdlife.org/europe-and-central-asia/news/exposed-slovenian-government-exploits-pandemic-shatter-environmental
29. Fashion Resale Market and Trend Report (2020) [Online]. ThredUP. Available at: https://www.thredup.com/resale/#resellable-brand. Accessed 17 Dec 2020
30. Fejerskov AM, Fetterer D (2020) Innovative responses to Covid-19: future pathways for 'techvelopment' and innovation. Danish Institute for International Studies, Copenhagen
31. Grevet A (2019) Vin suisse, des blancs et des rouges tout en précision [Online]. Vinoptimo. Available at: http://www.vinoptimo.com/news/negociant-en-vin/vin-suisse/. Accessed 12 Feb 2021
32. Hollanders H, Es-Sadki N, Merkelbach I (2019) Regional innovation scoreboard 2019. Publications Office of the European Union, Luxembourg
33. Kenny P (2019) "The enemy of the people": populists and press freedom. Polit Res Q 73(2)
34. Klonowska K, Bindt P (2020) The COVID-19 pandemic: two waves of technological responses in the European Union. Hague Centre for Strategic Studies, Hague
35. Kolowich Cox L (2021) The 18 best advertisements & ad campaigns of all time [Online]. Hubspot. Available at: https://blog.hubspot.com/marketing/best-advertisements. Accessed 30 Jan 2021
36. Kuciński K (2002) Geografia ekonomiczna: Zarys teoretyczny. Oficyna Wydawnicza Szkoły Głównej handlowej w Warszawie, Warsaw
37. Lab Grown Vs Mined (2020) [Online]. Clean origin. Available at: https://www.cleanorigin.com/about-lab-created-diamonds/. Accessed 13 Nov 2020
38. Liddiard P (2019) Is populism really a problem for democracy? [Online] Wilson Center's History and Public Policy Program. Available at: https://www.wilsoncenter.org/publication/populism-really-problem-for-democracy. Accessed 12 Feb 2021
39. Luhmann M, Hennecke M (2017) The motivational consequences of life satisfaction. Motiv Sci 3(1)

40. Markova T, Glazkova I, Zaborova E (2017) Quality issues of online distance learning. Proc Soc Behav Sci 237
41. Mazur J, Szkultecka-Debek M, Dzielska A, Drozd M, Malkowska-Szkutnik A (2018) What does the Cantril Ladder measure in adolescence? Arch Med Sci 14(1)
42. McAdam AJ et al (2021) When should asymptomatic persons be tested for COVID-19?. J Clin Microbiol 59(1)
43. Melamed D, Simpson B, Abernathy J (2020) The robustness of reciprocity: experimental evidence that each form of reciprocity is robust to the presence of other forms of reciprocity. Sci Adv 6(23)
44. Milasi S, González-Vázquez I, Fernández-Macías E (2020) Telework in the EU before and after the COVID-19: where we were, where we head to [Online]. Science for Policy Briefs of the European Commission. Available at: https://ec.europa.eu/jrc/sites/jrcsh/files/jrc120945_policy_brief_-_covid_and_telework_final.pdf
45. Morcos P (2020) Toward a new "lost decade"?: Covid-19 and defense spending in Europe [Online]. Center for Strategic and International Studies. Available at: https://www.csis.org/analysis/toward-new-lost-decade-covid-19-and-defense-spending-europe
46. Moy R (2020) Fashion retail after COVID-19 [Online]. Nextail. Available at: https://nextail.co/2020/04/07/fashion-retail-after-COVID-19/. Accessed 30 Jan 2021
47. O'Halloran J (2020) Global business leaders comfortable if not fully prepared for large-scale shift to remote work [Online]. Computer Weekly. Available at: https://www.computerweekly/news/252486320/Global-business-leaders-comfortable-if-not-fully-prepared-for-large-scale-shift-to-remote-work
48. Overview of Testing for SARS-CoV-2 (COVID-19) (2021) [Online]. Centers for Disease Control and Prevention. Available at: https://www.cdc.gov/coronavirus/2019-ncov/hcp/testing-overview.html. Accessed 12 Feb 2021
49. Pinckney J, Rivers M (2020) Sickness or silence: social movement adaptation to COVID-19. J Int Affairs 73(2)
50. Policy Brief 13: Interlinkages Between Energy and Jobs (2018) [Online]. International Renewable Energy Agency (IRENA), The European Commission and International Labour Organization (ILO). Available at: https://sustainabledevelopment.un.org/content/documents/17495PB13.pdf
51. Pozzer A, Dominici F, Haines A, Witt C, Münzel T, Lelieveld J (2020) Regional and global contributions of air pollution to risk of death from COVID-19. Cardiovasc Res 116(14)
52. Pył zawieszony: czym jest PM10, a czym PM2.5? Aerozole atmosferyczne (2018) [Online]. Airly. Available at: https://airly.eu/pl/pyl-zawieszony-czym-jest-pm10-a-czym-pm2-5-aerozole-atmosferyczne
53. Quality of life and quality of society during COVID-19 [Online]. Eurofound. Available at: https://www.eurofound.europa.eu/data/covid-19/quality-of-life. Accessed 12 Feb 2021
54. Quality of Life in European Cities (2013) [Online] Flash Eurobarometer 366 of the European Commission. Available at: https://ec.europa.eu/commfrontoffice/publicopinion/flash/fl_366_en.pdf
55. Regional Innovation Scoreboard 2019 Interactive Tool (2019) [Online]. EIS. Available at: https://interactivetool.eu/RIS/RIS_2.html?fbclid=IwAR3x7sGwrXlTeAXBnDkKKjWVnHn9Ug0_zPXMh4y-uNGRwppGo1qlZov5mks#a
56. Röth L, Afonso A, Spies DC (2018) The impact of populist radical right parties on socio-economic policies. Eur Polit Sci Rev 10(3)
57. Savills Global Luxury Retail 2019 Outlook (2019) [Online] Savills. Available at: http://pdf.savills.com/documents/global-luxury-retail-2019-outlook.pdf
58. Skok na polskie lasy!? (2020) [Online]. Fundacja Ekologiczne Forum Młodzieży. Available at: https://www.ekologiczneforummlodziezy.pl/pilne-skok-na-polskie-lasy/. Accessed 12 Feb 2021

59. Sprawozdanie ze stanu ochrony języka polskiego za lata 2016–2017. Język informacji politycznej (2019) [Online]. Rada Języka Polskiego przy Prezydium PAN. Available at: http://orka.sejm.gov.pl/Druki8ka.nsf/0/C4B224C28DB9367BC12583CB0032CA99/%24File/3324.pdf

60. Timbro Authoritarian Populism Index (2019) [Online]. Timbro. Available at: https://populismindex.com/. Accessed 12 Feb 2021

61. Trotter C (2020) What retailers need to know about resale in fashion [Online]. Insider Trends. Available at: https://www.insider-trends.com/what-retailers-need-to-know-about-resale-in-fashion/. Accessed 17 Dec 2020

62. Ucen P (2007) Parties, populism, and anti-establishment politics in East Central Europe. SAIS Rev 27(1)

63. Understanding Your GlobeSmart Profile (2019) [Online]. Globesmart. Available at: https://globesmart.zendesk.com/hc/en-us/articles/360033855213-Understanding-Your-GlobeSmart-Profile

64. Verick S, Islam I (2010) The great recession of 2008–2009; causes, consequences and policy responses. IZA discussion papers, no. 4934

65. World Economic Outlook Update January 2021 (2021) [Online]. International Monetary Fund. Available at: https://www.imf.org/en/Publications/WEO/weo-database/2020/October/. Accessed 12 Feb 2021

66. World Press Freedom Index [Online]. Reporters Without Borders (RSF). Available at: https://rsf.org/en/ranking. Accessed 12 Feb 2021

67. Zakon o spremembah in dopolnitvah Zakona o interventnih ukrepih za zajezitev epidemije COVID-19 in omilitev njenih posledic za državljane in gospodarstvo (ZIUZEOP-A) (2020) [Online]. Uradni list Republike Slovenije z dne 30.04.2020 (številka 61). Available at: https://www.uradni-list.si/glasilo-uradni-list-rs/vsebina/2020-01-0901/#1.%C2%A0%C4%8Dlen

68. Zipser D, Achille A (2020) A perspective for the luxury-goods industry during—and after—coronavirus [Online]. McKinsey. Available at: https://www.mckinsey.com/industries/retail/our-insights/a-perspective-for-the-luxury-goods-industry-during-and-after-coronavirus. Accessed 02 Feb 2021

Sustainable Management of Biomedical Waste During COVID-19 Pandemic

S. Grace Annapoorani

Abstract Biomedical waste is a hazard to worldwide communal environmental health. Productive biomedical waste administration is the need of great importance and a prompt order that needs to be addressed. Different stakeholders, for example, worldwide associations, government bodies, the scholarly community, biomedical waste treatment facilities and advancement suppliers are working inseparably to deal with the high amounts of biomedical waste created across the world. Appropriate biomedical waste (BMW) organization in understanding to the specified standard was one of the disregarded parts of medical care for quite a long time, particularly in non-developing nations like India. In the midst of the COVID infection 2019 (COVID-19) pandemic, the situation may deteriorate as confirmed by some underlying encounters, with heaps of personal protective equipment (PPE) collecting in the medical clinics. As is apparent from the current circumstance of the pandemic, essential disease control rehearses are the solitary measures for regulation. Appropriate garbage removal is a necessary piece of these control measures. There may be a genuine danger of spreading coronavirus whenever utilized covers, gloves, and other personal protective equipment are not handled and discarded appropriately. Also, household waste (e.g., tissues, face masks, gloves) puts waste management laborers at expanded health hazard. The COVID-19 pandemic and legislative strategies to contain the spread of virus have caused a universal financial downturn and have likewise produced a gigantic measure of biomedical waste. The arrangement was incredibly impacted by dispensable plastic-based personal protective equipment (PPE) and single-use plastics by online shopping for a large portion of the fundamental need. The utilization of PPEs and single-use plastics during the pandemic builds the amount of biomedical waste. Larger boundary of mobile services should be sustained, especially during the pandemic, which can be significant for the developing nations where the medical waste removal conveniences are restricted. The mobile services are helpful for the crisis circumstance as well as be utilized as an essential reinforcement limit.

S. G. Annapoorani (✉)
Department of Textiles and Apparel Design, Bharathiar University, Coimbatore,
Tamil Nadu 641046, India
e-mail: gracetad@buc.edu.in

Keywords Biomedical waste · Medical waste management · COVID-19 · Coronavirus

1   Introduction

COVID-19 is a pandemic that began in Wuhan, China and spread to the majority of countries. India is one of the countries with a large number of cases, and based on current trends, it may soon overtake the USA in terms of cases. Citizens, policymakers, physicians, and others will be affected for a long time as a result of the morbidity and mortality associated with COVID-19. As with other emerging diseases, there is currently no treatment or prevention choice that is focused on the concept of evidence-based medicine (EBM). As a result, a lot of research is being done all over the world to produce evidence for new COVID19 treatment modality BH ties and diagnostics. According to the article coronavirus research publishing: The rise and rise of COVID-19 clinical trials in January 2020, there were around 60 COVID-19-related clinical trials, which increased to around 4000 in the first week of July. A number of large and critical trials have already completed interim analysis and have published or are in the process of publishing their results. This is supposed to have a positive effect on patient care for COVID-19. Vaccines, which appear to be the only remedy for avoiding a second wave of infection and developing herd immunity, are currently in short supply in the world. As a result, scientists, researchers, and clinicians are likely to participate in clinical and basic research to identify treatment and diagnostic modalities for this disease in the near future.

To establish COVID-19 research priorities in India, the Indian Council of Medical Research National Task Force for COVID-19 established groups in clinical research, diagnostic and biomarker research, epidemiology and surveillance, operational research, vaccine and drug research. It is anticipated that high-quality research will be conducted in India as a result of these efforts at the government level and that the synthesized findings will be available to provide further guidance. While India ranks among the top ten countries in terms of COVID-19 research publications, no systematic review of registered clinical trials is available. The availability of such data is important not only for researchers and scholars, but also for policymakers and government officials to obtain a better understanding of the situation. With this goal in mind, we created this analysis to look into the data from the Clinical Trial Registry of India (CTRI) to see what kind of COVID 19 clinical trials are being performed in India. This will provide us with a snapshot of the current situation as well as details on future COVID 19 approaches from India [1].

2   Economic Consequences of the COVID-19 Outbreak: The Need for Epidemic Preparedness

COVID-19 has had a major impact on the global economy and financial markets, in addition to being a global pandemic and public health crisis. The disease control initiatives that have been introduced in many countries have resulted in substantial wage declines, increased unemployment, and disturbances in the transportation, service, and manufacturing sectors, to name a few. Most governments around the world tend to have underestimated the risks of rapid COVID-19 spread and have been largely reactive in their crisis response. Since disease outbreaks are unlikely to go away anytime soon, concerted international action is needed to save lives while also safeguarding economic prosperity.

2.1   COVID-19 and the Economy

COVID-19 was declared a pandemic by the World Health Organization (WHO) on March 11, 2020, citing over 3 million cases and 207,973 deaths in 213 countries and territories the report was published according to World Health Organization Coronavirus Disease 2019 (COVID-19): Situation Report 100. (2020). The virus has not only been a public health emergency, but it has also had an economic impact on the world. Reduced productivity, loss of life, business closures, trade disruption, and the annihilation of the tourism industry have all had significant economic consequences around the world. COVID-19 may be a "wake-up" call for world leaders to step up collaboration on disease preparedness and provide the funding needed for global collective action. While there is ample knowledge on the potential economic and health costs of infectious disease outbreaks, the world has failed to invest sufficiently in prevention and preparedness measures to reduce the risks of major epidemics.

Infectious disease outbreaks and epidemics have become global challenges as a result of globalization, urbanization, and environmental change, necessitating a concerted response. Despite the fact that the majority of developed countries, mostly those in Europe and North America, have strong real-time surveillance and health systems to manage infectious disease spread, improvements in public health capacity in low-income and high-risk countries—including human and animal surveillance, workforce preparedness, and laboratory resource strengthening—must be backed up by national resources. Building and funding technical platforms to accelerate the research and development response to new pathogens with epidemic potential has been promoted by governments, non-governmental organizations, and private companies on a global scale.

Larger economic issues are linked to current and expected future oil demand, resulting in price increases as a result of decreased economic activity caused by the COVID-19 pandemic. Large price declines were also triggered by expected surplus supply. Migrant workers contribute significantly to global labor markets, addressing inequalities in both high- and low-skilled jobs. As countries try to halt the spread of COVID-19, international travel restrictions and quarantine are likely to stay in place for the near future; migration flows will be limited, limiting global economic growth and development [2].

2.2 Impacts of Coronavirus on Solid Waste Management

The COVID-19 outbreak will generate controlled medical waste, which may include needles, sharps, material contaminated with bodily fluids (such as gauze, gloves, or gowns), and pathological wastes. The Occupational Safety and Health Administration (OSHA) and the Department of Transportation (DOT) oversee these wastes on a federal level for proper handling and transportation, as well as at the state level for control and care of these materials before disposal.

The risk of contracting COVID-19 is greater for individuals who are in close contact with someone who already has the disease, according to the US Centers for Disease Control and Prevention (CDC). When an infected individual coughs or sneezes, the virus is thought to spread primarily by respiratory droplets (not truly airborne). It's also possible that a person could contract COVID-19 by touching a virus-infected surface or object and then touching their mouth, nose, or eyes, but this isn't thought to be the main way the virus spreads.

Medical waste produced in the care of COVID-19 patients and patients under investigation (PUIs) should be handled according to standard procedures, according to the CDC. Waste produced in the treatment of PUIs or patients with confirmed COVID-19 does not require additional considerations for wastewater disinfection in the USA, according to the CDC. Coronaviruses are vulnerable to the same disinfection conditions as other viruses in community and healthcare environments, therefore existing disinfection conditions in wastewater treatment facilities should be adequate.

2.3 Guidelines for Handling of Waste Generated During COVID-19 Patient's Treatment

The guidelines are focused on current COVID-19 awareness and current practices in the management of infectious waste produced in hospitals while treating viral and other contagious diseases such as HIV, H1N1, and others. Healthcare facilities with isolation wards for COVID-19 patients must hold separate color-coded bins/bags/containers in the wards and ensure proper waste segregation, according to the guidelines. To ensure sufficient strength and no leaks, double-layered bags should be used for waste collection from COVID-19 isolation wards. Separately collect and store biomedical waste before delivering it to the Popular Biomedical Waste Treatment and Disposal Facility (CBWTF). To store COVID-19 waste, use a dedicated collection been labeled "COVID-19" and keep it separate in a temporary storage room before handing it over to CBWTF approved workers. Biomedical waste collected in isolation wards can be transferred directly from the ward to the CBWTF collection van.

Bags/containers used for collecting biomedical waste from COVID-19 wards should be labeled as "COVID-19 Waste," in addition to the required labeling, it said, adding that general waste that is not contaminated should be disposed of as solid waste according to the Solid Waste Management Rules, 2016. Keep track of the waste produced in the COVID-19 isolation wards in a separate file. In COVID-19 isolation wards, use dedicated trolleys and storage bins. These products should also have a mark that says "COVID-19 Waste." The (inner and outer) surfaces of containers/bins/trolleys used for COVID19 waste storage should be disinfected with a 1% sodium hypochlorite solution on a regular basis [3].

3 Biomedical Waste Management: The Challenge During COVID-19 Pandemic

For years, proper biomedical waste (BMW) management in compliance with the rules has been one of the most overlooked aspects of health care, especially in developing countries like India. After the Government of India's (GoI) Ministry of Environment, Land, and Climate Change implemented reforms in 2016 by prescribing simpler categories (color-coded) for segregation of different BMWs, an amendment in 2018 came into effect with the aim of enhancing rule enforcement. According to the article Biomedical waste management in India: awareness and novel approaches, still, with an annual growth rate of 7% and a projected estimate of 775.5 tonnes/d by 2022, proper segregation, storage, and disposal remained a serious concern for healthcare facilities across India.

As demonstrated by some early experiences, the situation could escalate during the coronavirus disease 2019 (COVID-19) pandemic, with piles of personal protective equipment (PPE) accumulating in hospitals. Despite World Health Organization and Ministry of Health and Family Welfare, Government of India guidelines on the fair use of personal protective equipment (PPE) for COVID-19, healthcare settings are experiencing high demand for PPE from all strata of healthcare staff due to fear of infection.

The fear also leads to the misuse of PPE, exacerbating the issue by causing a large number of BMWs to be produced, which are difficult to store and transport with the limited resources and manpower available during a crisis. The indiscriminate use and disposal of single-use surgical masks (at times N95 respirators) also in the community adds to the danger. Their disposal is often mixed with non-infectious kitchen/general waste from households and residential areas where color-coded BMW bins are hard to come by. It's worth noting that the indiscriminate dumping of BMWs in the general garbage provides quick access to the largely susceptible population for the current severe acute respiratory syndrome coronavirus (SARS CoV2) with fomite-borne transmission and a yet-to-be-determined time of viability on fomites.

In response to the COVID-19 pandemic in India, the Ministry of Environment, Forest & Climate's Central Pollution Control Board (CPCB) has issued guidelines for the management of waste produced during COVID-19 patient care, diagnosis, and quarantine. In addition to the recommendation to adopt current practices of BMW Management Rules, 2016, these guidelines advocated for the use of double-layered bags (using two bags), mandatory marking of bags and containers as "COVID-19 waste," routine disinfection of dedicated trolleys, and separate record keeping of waste produced from COVID-19 isolation wards. Although the CPCB is the nodal agency in India for any BMW-related recommendations, other government agencies have issued guidelines for the management of COVID-19 waste. Although the above guidelines must follow the CPCB's recommendations, there have been some discrepancies in the segregation norms of the current BMW Management Rules, 2016 in which the items are separated based on the final mode of care. This sudden shift in the category of some BMWs may lead to inadvertent final care.

In the upcoming CPCB guidelines, spill management from BMW bags or containers, recommendation of ideal PPE for persons handling BMW, and provision of health check-ups for those individuals may be considered. Other government agencies should make it clear in their own guidelines that the CPCB's standards for BMW management should be followed, with a link to the CPCB's guidelines. Monitoring the execution of the prescribed policies for BMW management should be prioritized, in addition to ensuring uniformity in the guidelines.

Due to an increase in the amount of waste, hospitals and institutions should increase their capacity to transport and store BMWs. An electronic correspondence system should be in place to determine the exact amount of "COVID-19 waste" in various categories, as well as any breaches during transportation and treatment that should be reported to the nodal department. BMW management should be educated on a regular basis by the institution's infection prevention and control unit, which should also oversee the activities. These measures, when combined with strict adherence to the laws, would ensure that the imminent crisis is managed more effectively. Basic infection prevention practices are the only mechanisms for containment, as shown by the latest pandemic situation. The proper management of waste is an important part of these control steps. The authors hope that by sending this letter, they will be able to draw the attention of concerned authorities at the institutional and local administrative levels to the importance of implementing the prescribed policies in order to improve overall BMW segregation and disposal in daily practice, which will be of great assistance even after the pandemic [4].

4 COVID-19 and Readjusting Clinical Trials

Clinical studies have been delayed around the world as a result of the COVID-19 pandemic, with long-term consequences for medical research. Clinical trial activity has been severely disrupted around the world as a result of the COVID-19 pandemic. The virus has had a significant impact on the ability to perform trials in a healthy and successful manner, much as it has in other areas of life. This is particularly true given that clinical trials often include vulnerable populations who are most at risk from COVID-19 exposure. Thousands of trials have been postponed or stopped due to the difficulty of continuing them under lockout, despite the fact that the controls have started to ease in some parts of the world. Simultaneously, the pandemic has resulted in an extraordinary shift in clinical trial research toward COVID-19.

Clinical trials have long been the gold standard for assessing and validating experimental treatments and therapies. Effective studies into the safety and effectiveness of new drugs are needed for new drug approval. Hundreds of different sites around the world can be involved in trials, each with its own set of circumstances, results, and government rules on what is permissible. When you realize how many participants are participating in a clinical trial, the complexity of the issue becomes overwhelming. Clinical trials include clinical caregivers and nurses who work with patients at clinical trial sites, postgraduate researchers, postdoctoral fellows, research scientists, and others who work on the trial protocol and secure funding (either from governments, foundations, pharmaceutical or device manufacturers, or a combination of the above). Clinical trials often include postgraduate researchers, postdoctoral fellows, research scientists, and others who work on the trial protocol and secure funding (either from governments, foundations, pharmaceutical or device manufacturers, or a combination of the final result).

Clinical trials are an important tool in medical science, but COVID-19 has revealed several ways to enhance their design, behavior, and reporting. The rapid design and launch of COVID-19 clinical trials have shown that some elements and procedures of clinical trials could be strengthened, simplified, or modernized in ways that would favor patients, physicians, and all science. In order to ensure the highest standard of research in the future, it will be crucial to integrate those lessons into the research process [5].

5 Classification of Biomedical Waste

5.1 Non-Hazardous Waste

The majority of the waste produced by healthcare facilities is non-hazardous waste (approximately 85%). It consists of food scraps, paper cartons, wrapping materials, fruit peels, and washes water, among other things.

5.2 Hazardous Waste

(A) **Potentially contagious waste**: Various occasions for infectious waste have been used in scientific literature, guideline guides, and codes over the years. Infectious medical waste, toxic, red container, and infected medical waste, as well as controlled and non-regulated medical waste, are all examples. Both of these terms apply to the same kind of waste, but the terms used in legislation are generally more descriptive. It accounts for 10% of total waste, which includes the following items:

1. Dressings and swabs infected with blood, pus and body fluids.
2. Laboratory waste including laboratory culture stocks of infectious agents
3. Potentially infected material: Excised tumors and organs, placenta.
4. Potentially infected animals used in diagnostic and research studies.
5. Sharps, which include needle, syringes, blades, etc.
6. Blood and blood products.

(B) **Potentially toxic waste**

1. Radioactive waste: It includes waste contaminated with radionuclide; it may waste generated from in vitro analysis of body fluids and tissue.
2. Chemical waste: It includes disinfectants, X-ray processing solutions, monomers and associated reagents, base metal debris (dental amalgam in extracted teeth).
3. Pharmaceutical waste: It includes anesthetics, sedatives, antibiotics, analgesics, etc.

5.3 Steps for Effective B.M.W Management

Waste Survey

It is an important component of the waste management method. A survey helps in assessment of both the type and amount of waste generated.

- Waste survey is valuable in the aspect of:
- Make a distinction of types of waste.
- Enumerate the waste generated.
- Conclude the points of generation and type of waste generated at each point.
- Determine the altitude of generation and disinfection within the hospital.
- To trace out the type of disposal carried out.

5.4 Segregation

Segregation is the primary division of various types of waste produced on a daily basis, lowering the risks as well as the cost of handling and disposal. In the management of biomedical waste, segregation is the most important step. Only effective segregation will ensure that biomedical waste is properly managed.

The BMWs must be separated according to the guidelines outlined in Schedule 1 of the BMW Rules, 1998.

At the point of generation, various types of waste are placed in different containers or coded bags. It contributes to the reduction of infectious waste and treatment costs. Segregation also tends to prevent the virus from spreading and decreases the risk of infecting other healthcare workers.

5.5 Treatment of Waste

Treatment refers to any process that alters waste in any way before it reaches its final destination. It is primarily essential to disinfect or decontaminate the waste at the source so that pathogenic organisms are no longer present. Following this treatment, the remains can be securely treated, transported, and stored.

- In syringe cutters and needle destroyers, syringe nozzles and needles should be shredded.
- Broken glass, scalpel blades, and Lancets should be kept in separate containers of bleach, then moved to plastic/cardboard boxes and sealed to avoid spillage before being shipped to incubators. Glassware must be sterilized, disinfected, and washed.
- Culture plates with possible culture must be autoclaved; media are placed in suitable bags and disposed off. The plates can be reused after sterilization.
- Gloves should be cut/shredded/mutilated before disposal.
- Swabs should be chemically sterilized followed by incineration. If they have only a small quantity of blood that does not drip, they could be placed in the garbage.
- Disposable objects should be dipped in freshly prepared 1% sodium hypochlorite for 30 min to an hour, after which mutilation is needed before disposal, according to the policy.
- Under no conditions can heat be used to dispose of amalgam. Mercury can volatize and be released into the atmosphere as a result. As a result, amalgam-filled teeth should be immersed in high-level disinfectants (e.g., Gluteraldehyde) for 30 min.
- The laboratory's liquid waste is either pathological or chemical. Reagents can be used to defuse non-infectious waste.
- Liquid contaminated waste should be treated with a chemical decontaminator and then defused.

5.6 *Minimization of Waste*

Although standard solid and liquid waste does not need to be handled before being disposed of, nearly all infectious waste should be treated first. The cost of disposing of infectious waste may be ten times that of ordinary waste disposal. Any approach that reduces the amount of infectious waste produced would also reduce the cost of infectious waste disposal.

6 Management of Plastic in Healthcare

Plastic is used in health care in the form of disposable syringes, blood, bottles, and urine bags, surgical gloves, catheters, and other products. Plastic has been linked to a drop in sperm count, genital abnormalities, and an increase in breast cancer cases. Carcinogens such as dioxin and furan are released when plastics are burned. Because of its non-biodegradable nature, plastic has become a serious environmental and health issue. Land filling or recycling in the long run. Before returning to the vendor, all disposable plastic should be shredded. The development of environmentally friendly, biodegradable plastics is urgently needed. It is also critical to reduce the amount of plastic waste generated [6].

6.1 *Characterization of Healthcare Waste*

Healthcare waste includes all waste created inside healthcare facilities, research centers, and laboratories related to medical procedures, according to WHO guideline reports. It also includes waste created by healthcare providers in their homes (e.g., home dialysis, self-administration of insulin, recuperative care). There are eight main categories of healthcare waste, which include both hazardous and non-hazardous materials.

6.2 *Infectious Waste*

Waste contaminated with blood and other body fluids (e.g., from discarded diagnostic samples), infectious agent cultures and stocks from laboratory work (e.g., waste from autopsies and infected animals from laboratories), or waste from infected patients (e.g., swabs, bandages and disposable medical devices).

6.3   Pathological Waste

Human tissues, organs or fluids, body parts and contaminated animal carcasses. Sharps waste Syringes, needles, disposable scalpels and blades, etc.

6.4   Chemical Waste

Solvents and reagents used for laboratory preparations, disinfectants, sterility and heavy metals contained in medical devices (e.g., mercury in broken thermometers) and batteries.

6.5   Cyctotoxic Waste

Waste containing substances with geno-toxic properties (i.e., highly hazardous substances that are, mutagenic, teratogenic or carcinogenic), such as cytotoxic drugs used in cancer treatment and their metabolites.

6.6   Radioactive Waste

Products contaminated by radio nuclides including radioactive diagnostic material or radio-therapeutic materials.

6.7   Pharmaceutical Waste

Pharmaceutical Waste is like Expired, unused and contaminated drugs and vaccines.

7   Non-Hazardous or General Waste

Waste that does not pose any particular biological, chemical, radioactive or physical hazard. The Compendium on Technologies for the Treatment/Destruction of Healthcare Waste (UNEP-IETC 2012) also includes baseline data on the composition and volume of healthcare waste, as well as potentially infectious contents. The waste produced in healthcare facilities is typically estimated to be 85%

non-hazardous and 15% hazardous. Baseline data on the composition and amount of healthcare waste, as well as potentially infectious contents are included in the Compendium. In most healthcare facilities, waste is calculated to be 85% non-hazardous and 15% hazardous.

In addition to material constituents, knowledge of the properties of healthcare waste is needed in order to properly select acceptable options for handling healthcare waste, define treatment technologies, and set necessary criteria for treatment system operation. Based on UNEP-IETC data, the moisture content, heating value, percentage of combustible materials, and bulk densities of healthcare waste are summarized in general conditions (2012). However, in the current situation, due to the COVID-19 pandemic, these parameters could change.

8   Volume of Healthcare Waste Generation

Under normal circumstances, the overall healthcare waste generation rates by form of medical facility. The highest generation of healthcare waste, according to this data, occurs in maternity centers and hospitals.

In terms of healthcare waste generation in developing countries during the COVID-19 pandemic, there could be a rise in healthcare waste volumes in five Asian cities (ADB 2020). The rise in healthcare waste from healthcare facilities associated with COVID-19 is projected to be 3.4 kg/person/day shown in Fig. 1.

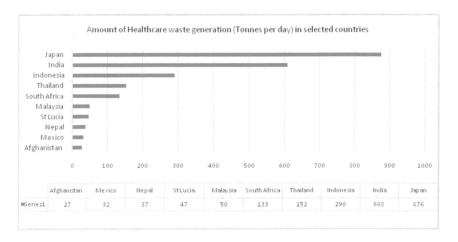

Fig. 1 Amount of healthcare waste generation in selected countries

9 Policy and Regulatory Aspects

The use of current healthcare waste management plans and strategies by national, regional, and local governments in their response to COVID-19 waste would help them greatly. Current disaster waste contingency plans, especially those that include healthcare waste, can be useful in the context of COVID-19 in the municipal solid waste field. When designing strategies, plans, or policies for municipalities or countries that do not yet have them, they should provide contingency planning for disease circumstances, and this material can be guided by ongoing local COVID-19 waste management challenges.

In addition to current national policies and plans, most of the countries that participated in this study's questionnaire survey provided policies, guidelines, and plans to respond to the COVID-19 pandemic right away. In times of COVID-19, as in any emergency, it's vital to have a specific assignment of responsibility, as well as adequate human and material resources to safely dispose of COVID-19-related waste.

According to the information provided above, many developing countries face difficulties not only in developing policies and issuing guidelines, but also in enforcing them once they have been prepared. As a result, a national healthcare waste management strategy is expected to guide political decision-making and coordinate government actions and resources to ensure that the programs are implemented successfully (WHO 2014; UNEP 2020). The following are some of the main factors to remember when developing healthcare waste management policies:

- **Identification of needs and gaps in the country**, considering international agreements and conventions adopted nationally, sustainable development and environment and safe management of hazardous waste.
- **Regulations** specifying HCW management such as waste segregation, collection, storage, handling, treatment, disposal and transport of different waste categories, responsibilities, and training requirements.
- **Practical/technical guidelines** and manuals supplementing the official regulations on implementation, directly applicable to local managers and staff.
- **Regularly updated allocation of roles**, resources, and responsibilities, describing actions to be implemented by authorities, healthcare personnel, and waste workers.
- **A contingency planor** to ensure the continuity of healthcare waste management systems during an emergency. Since resources can be scarce in an emergency, contingency plans should provide alternate options for staff, vehicles, infectious waste, waste accumulation, washing, disinfection, and street cleaning. Furthermore, MSW management is a vital service that must be accessible at all times during a pandemic. Disrupting MSW management services on a regular basis can result in additional social and public health problems, which must be avoided. COVID-19 waste created in homes and public places can be handled in

Table 1 Hierarchy of the regulatory and Institutional framework for healthcare waste management

International Laws and Conventions Regulations and Guidelines	National level institution • National government (ministries, agencies) • National (ad-hoc) taskforces
National law	Provincial level institution • Provincial government (agencies) • Provincial (ad-hoc) taskforces
National healthcare waste management policy • National healthcare waste regulations and guidelines • National healthcare waste management plan	Municipal level institution • Provincial government (agencies) • Provincial (ad-hoc) taskforces
Local healthcare waste management regulations and guidelines • Local healthcare waste management plan	Institutional level—healthcare facilities
Institutional healthcare waste management plan (healthcare facilities level)	–

compliance with current policies, legislation, strategies, and plans. In some countries, specific warnings and operations for such waste have been developed (Table 1).

10   COVID-19 and Gender in Waste Management

There is still a shortage of gender-disaggregated waste and COVID-19 info. However, there are a number of underlying factors that indicate the COVID-19 pandemic could place women at greater risk in terms of their health, social, and economic well-being, as well as worsen current gender disparities. Gender inequality, accountability, and responsibilities are profoundly rooted in certain areas of waste management, despite the fact that it is often considered gender-neutral. As a consequence, in the sense of COVID-19, it is important to consider the relationship between gender and waste.

An estimated 20 million people around the world depend on informal waste recycling to make a living. Women and children, for example, are an important part of the informal waste sector in Asia, where they make a living through waste collection and/or recycling. Indeed, the study found that in Bhutan, Mongolia, and Nepal, more women than men work in the informal sector, where they often perform tasks with little to no safety equipment and for low pay, such as waste picking at landfills. As a result of the COVID-19 pandemic, when infectious waste is discarded at disposal sites, women could be placed at risk. While cultural factors which exclude women from decision-making and limit their access to information on outbreaks and service availability, it is critical that women have access to accurate information on how to use precautionary measures such as personal protective equipment (PPE), as well as health services and insurance to ensure their health and safety.

This also highlights how, in the informal sector, women are mostly relegated to lower-paying jobs like waste picking, sweeping, and waste separation, while men may hold positions of higher authority, such as dealing with the buying and reselling of recyclables. Women are traditionally removed from higher-paying and decision-making positions in the formal waste economy as well. If the COVID-19 pandemic persists, it may pose a significant economic threat to women working in the waste industry, widening gender disparities in livelihoods.

This also suggests that current waste management policies and standards are not gender-specific. Despite the fact that women are typically the primary handlers of household waste and that the waste sector has a strong gender division of labor, many guidelines fail to discuss particular needs and issues related to gender. Gender mainstreaming in the waste sector may be a way to develop waste management in a more resilient and long-term way. Households, for example, have considerable collective potential to minimize waste flow into waste management systems while having the least organized involvement in the waste sector's guidance and policy frameworks. Women and men will play an important role not only in accelerating waste reduction, segregation, composting, and recycling in general, but also in ensuring protection during the COVID-19 pandemic by properly segregating waste at the household level.

Collection of gender-disaggregated data: Gender-disaggregated data on waste and the COVID-19 outbreak will assist in identifying gendered disparities in infectious waste exposure and designing gender-responsive COVID-19 pandemic regulations and guidelines.

Women's representation: Encouraging women to participate in decision-making and the number of women in decision-making roles in waste management will help to ensure that women have a say in important waste management decisions.

Health and safety: Women may be excluded from decision-making and have limited access to information on outbreaks and service availability due to cultural factors. As a result, it's important that women have access to reliable knowledge on how to use precautionary procedures like wearing protective equipment while treating healthcare and infectious waste, as well as health facilities and insurance to ensure their health and safety.

10.1 Waste Segregation, Storage, and Transportation of COVID-19 Waste

Healthcare waste must be adequately isolated at the source, preserved, and transported not only to avoid negative health and environmental effects, but also to ensure resource quality and material recovery. Furthermore, current operating guidelines for healthcare waste management and MSW management should be followed for COVID-19 waste, with specific precautionary steps, modifications, and arrangements implemented to minimize any possible risks of COVID-19 contamination during the waste management process.

10.2 Waste Minimization

Better practice in healthcare waste management should try to prevent or recover as much waste as possible, rather than disposing of it by burning or burial, according to the waste management hierarchy based on the definition of the 3.Rs (reduce, reuse, and recycle). The best waste management approach is to stop producing waste in the first place by eliminating inefficient working practices. While waste minimization is typically practiced at the point of generation, such as the separation of hazardous waste from other wastes, a well-thought-out strategy that combines buying and stock management strategies can also reduce the amount of waste generated.

10.3 Waste Segregation at Source

Color-coding makes it easier for waste-handling workers to position waste products in the required container and maintain waste segregation during transportation, storage, care, and disposal. Color-coding also gives a visual indication of the danger that the waste in a given container poses. Many countries have national legislation that defines the forms of waste to be segregated and a color-coding scheme for waste containers. A World Health Organization (WHO) scheme is available where there is no national legislation. Each country has issued some specific regulations, operations and recommendations for proper waste segregation and handling at source that includes COVID-19 waste generated by healthcare facilities, as shown in Fig. 2.

- Segregate waste as close to the source as possible (proximity principle).
- Place segregated waste in identifiable, color-coded, labeled containers or bags, which are leak-proof and puncture resistant (particularly for sharps).

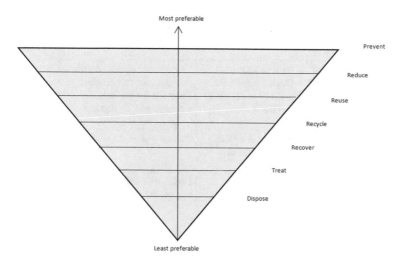

Fig. 2 COVID 19 waste generated by Healthcare facilities

- Place instructions for proper waste segregation close to the container.
- Use double-layer bags. Waste is to be placed in a specialized bag or container, sealed, and then placed in the second bag or container.

10.4   In-House Transport and Storage

Where feasible, onsite transportation can take place at less busy hours. Set routes should be followed to keep staff and patients safe and to reduce the passage of filled carts through clean areas including patient care. Transport routes and collection times should be set in stone and dependable. Infectious or other hazardous waste should not be collected at the same time or with the same trolley as general waste. Storage should be held away from patients and the general public. It should also be well ventilated and out of reach of vertebrate pests (Table 2).

10.5   Transportation to Offsite Treatment

Offsite transportation refers to activities such as transporting medical waste from healthcare facilities to treatment facilities via public roads. For designated COVID-19 healthcare facilities, it is critical to provide routine and expanded waste collection services. When transporting healthcare waste, use specialized, approved healthcare waste service providers wherever possible. Hazardous wastes should be transported through international borders for treatment in accordance with national

Table 2 WHO recommended segregation and collection scheme

Waste categories	Color of container and markings	Type of container	Collection frequency
Infectious waste	Yellow with biohazard symbol (highly infectious waste should be additionally marked HIGHLY INFECTIOUS)	Leak- proof strong plastic bag placed in a container (bags for highly infectious waste should be capable of being autoclaved)	When three-quarters filled or at least once a day
Sharp waste	Yellow, marked SHARPS with biohazard symbol	Puncture-proof container	When filled to the line of three- quarters
Pathological waste	Yellow with biohazard symbol	Leak-proof strong plastic bag placed in a container	When three-quarters filled or at least once a day
Chemical and pharmaceutical waste	Brown labeled with appropriate hazard symbol	Plastic or rigid container	On demand
Radioactive waste	Labeled with radiation symbol	Lead box	On demand
General healthcare waste	Black	Plastic bag inside a container which is disinfected after use	When three-quarters filled or at least once a day

Source Module 2: The Healthcare Waste Management System (https://www.who.int/topics/medical_waste/en)

laws and international agreements (Secretariat of the Basel Convention 1992). In the absence of national regulations, responsible authorities can depend on UN recommendations on the transportation of dangerous goods. Before sending hazardous healthcare waste offsite, some countries have implemented a manifest system or waste-tracking notice that includes the following information: (i) waste sort, (ii) waste sources, (iii) pick-up date, (iv) destination, (v) driver's name, (vi) number of containers or volume, and (vii) receipt of load from responsible person at the pick-up location.

While the majority of countries that responded to the questionnaire survey took concrete measures on healthcare waste management from healthcare facilities, as mentioned above, only a few implemented legislation, operations, and recommendations for segregation, storage, and transportation of COVID-19 waste created from households and quarantine locations (Table 3).

As a result, household healthcare waste, such as potentially contaminated masks, gloves, and discarded or expired medicine, should be classified as infectious waste, disposed of separately, and processed by municipal professionals or private waste management companies.

Table 3 Existing practices of storage and transport of healthcare waste in selected countries

Country	COVID-19 waste generated from a household or a quarantine location
Afghanistan	• Carried out no waste separation at source
Kenya	• Treated as potentially infectious and disinfect • Sealed the bag when its 2/3 full, with appropriate adhesive tape • Segregated and stored in leak-proof liner bags/container labeled "infectious waste"
Mexico	• Designated the temporary storage for waste biological- infectious hazardous • Stored packaged biological- infectious hazardous in metal or plastic containers with a lid and a label • Defined the temporary storage period according to the HCW types
South Africa	• Treated the waste such as disposable cloths, tissues, gloves, and masks generated from possible cases • Cleaned HCW areas • Placed red liners in a second red liners and tied • Placed in a suitable and secure place and marked or storage until the individuals test results are known • Avoided from placing HCW in communal waste areas • Stored the bags separately for five days in the sun before they are put out for collection by the municipality
India	• Use dedicated carts/trolleys/vehicles for transport of biomedical waste • Ensure sanitation of vehicles with 1% sodium hypochlorite after each trip

Source Module 2: The Healthcare Waste Management System (https://www.who.int/topics/medical_waste/en)

10.6 Mapping Source of Waste Generation in Order to Identify Changes in Waste Amounts and Flows and Increase Resource Use Efficiency

• Map outsources generating potentially COVID-19—contaminated waste, and healthcare waste, including hospitals, homecare centers, testing labs, quarantine camps and homes with self-isolated patients (if legally permitted).
• Map outsources where wastewater generation has decreased due to preventive closures, which may include schools, commercial complexes, and public places.
• Identify places where illegal dumping is taking place.

10.7 Separate Contaminated Waste from Households

• All potentially infectious waste should be placed in a sealed bag and, if necessary, double bagged. It must be treated as a residual waste that is not intended for material recovery.

- If it is difficult to distinguish contaminated waste from other household waste, all waste from that household should be put in a double-layered bag and sealed.
- Local governments should recommend providing waste bags to households and communities if necessary.

10.8   Minimization and Recycling

- Conduct waste minimization.
- Ensure continuity of recycling, so that the household keeps on segregating recyclables and non-recyclables. In general, there are three types of waste separation recommended:

(i) Mixed waste (including special waste as seen in Sri Lanka, for example): (ii) Recyclables: (iii) Organic waste.

10.9   Packaging and Storage

- Mixed waste and recyclables should be kept prior to collection.
- Mixed waste and special waste that contains potentially infectious materials should be double bagged.
- Some municipalities or private collection services provide special colored bags with certain thickness requirement specifically for collection of waste containing potentially contaminated materials.

10.9.1   Municipal Waste Collection and Transport

- Based on mapping of evolving waste sources and flows, human and financial resources and assets for waste management can be reassigned.
- In communities with common collection points and high population density, particularly informal settlements, the frequency and coverage of waste collection can be increased.
- Collections from the informal sector should be actively engaged in solving the crisis.
- It is possible to consider ad-hoc facilities, such as those provided by third-party operators, for the collection of waste from COVID-19 households in separate containers.
- The frequency at which recyclables are collected may be updated.

11   Treatment and Disposal Methods of Healthcare Waste

Healthcare waste, particularly COVID-19 waste, must be handled according to local guidelines and regulations, which usually involve thermal treatment. The pros and cons of some of the most popular technologies for the treatment and destruction of healthcare waste, as well as a sustainable evaluation of technology for choosing among them, are discussed below.

12   Incinerator

Incineration is a high-temperature, dry oxidation process that converts organic and combustible waste to inorganic, incombustible matter and decreases waste volume and weight significantly. High-heat thermal processes occur at temperatures ranging from 200 °C to over 1000 °C. Combustion, pyrolysis, and gasification are all processes that require the chemical and physical breakdown of organic material. However, the Stockholm Convention has proposed using the Best Available Technologies (BAT) with a combination of appropriate primary and secondary steps to monitor dioxin and furan air emissions no higher than 0.1 ng I-TEQ3/Nm10 (at 11% O_2) and less than 0.1 ng I-TEQ/l for wastewaters discharged from the facility to control environmental pollutions (UNEP 2007) (Table 4).

13   Autoclave

Autoclaves have been used to sterilize surgical instruments for over a century, and they have recently been modified for the treatment of infectious waste. An autoclave is made up of a metal vessel designed to withstand high pressures, a sealed door, and a system of pipes and valves that allow steam to enter and exit the vessel (Table 5).

Table 4 Pros and cons in applying the incineration option

Pros	Cons
• Significant reduction of waste	• High energy requirement
Volume and weight	• The combustion of healthcare waste produces mainly gaseous emissions, including steam, carbon dioxide
• Ensured contamination (Combustion at minimum 800 °C temperature)	• Nitrogen oxides, a range of volatile substances (e.g., metals, halogen acids, products of incomplete combustion)
	• Potential emissions of carcinogens
• No post treatment needed for final disposal	• Particulate matter, plus solid residues in the form of ashes, which are to be treated as toxic

Source Health Hazards of Medical Waste and its Disposal (https://www.nih.gov)

Table 5 Pros and Cons in applying the autoclave option

Pros	Cons
• Suitable for soiled wastes, bedding and personal, protective equipment, clinical laboratory waste, reusable instruments, waste sharps, and glassware • Low-heat thermal processes produce significantly less air pollution emissions than high-heat thermal processes • No specific pollutant emissions limits for autoclaves and other steam treatment systems • Waste does not require further processing, it can be disposed on a municipal landfill as it is disinfected and not hazardous anymore. However, some countries request to render the waste unrecognizable then it is shredded afterward, but this depends on the legal regulation • Available in various sizes from lab autoclaves to large autoclaves used in large waste treatment facilities	• Cannot treat volatile and semi-volatile organic compounds, chemotherapeutic waste, mercury, other hazardous chemical and radiological waste, large and bulky bedding material, large animal carcasses, sealed heat-resistant containers • Odors can be a problem around autoclaves if their insufficient ventilation • Poorly segregated waste may emit low levels of alcohols, phenols, formaldehyde, and other organic compounds into the air • Treated waste from an autoclave retains its physical appearance • Waste requires further processing for final disposal

Source Health Hazards of Medical Waste and its Disposal (https://www.nih.gov)

14 Microwave Treatment

Microwave technology is basically a steam-based method in which moist heat and steam produced by microwave energy are used to treat patients. Microwave energy with a frequency of about 2450 MHz and a wavelength of 12.24 cm rapidly heats water in the waste (Table 6).

15 Disposal

The majority of residual waste left after each of the above treatment options is disposed of on soil. If necessary, this should be achieved in a managed or sanitary landfill. Most countries have made some efforts to implement applicable technologies to handle healthcare waste created by healthcare facilities, according to country-based data [7].

Table 6 Pros and Cons in applying the microwave option

Pros	Cons
• Suitable for soiled wastes, bedding and personal, protective equipment, clinical laboratory waste, reusable instruments, waste sharps, and glassware • A fully enclosed microwave unit can be installed in an open area, and used with a HEPA filter to prevent the release of aerosols during the feed process • A large-scale, semi-continuous microwave unit is capable of treating about 250 kg/hour (3,000 tonnes per year) • Waste does not require further processing, it can be disposed on a municipal landfill as it is disinfected and not hazardous anymore. However, some countries request to render the waste unrecognizable then it is shredded afterward, but this depends on the legal regulations • Available in various sizes from lab autoclaves to large autoclaves used in large waste treatment facilities	• Volatile and semi-volatile organic compounds, chemotherapeutic waste mercury, other hazardous chemical and radiological waste, large and bulky bedding material, large animal carcasses, sealed heat-resistant containers • Treated waste from an autoclave retains its physical appearance • Waste requires further processing for final disposal • Very limited volume reduction, no weight reduction

Source Health Hazards of Medical Waste and its Disposal (https://www.nih.gov)

15.1 Handling Healthcare Waste Generated During the COVID-19

- **There is limited availability and accessibility of treatment and disposal options for healthcare waste management in developing countries.** This caused difficulty in handling healthcare waste generated during the COVID-19. In addition, the delivery of healthcare waste treatment and disposal facilities is not standardized across countries. The majority of treatment facilities are concentrated in urban areas, leaving peri-urban and rural areas with limited treatment options. In India, some states, including Arunachal Pradesh, Andaman and Nicobar, Goa, Lakshadweep, Mizoram, Nagaland, and Sikkim, lack Common Biomedical Waste Treatment and Disposal Facilities for the treatment and disposal of biomedical waste.
- **Another common problem is that, due to a lack of expertise, capacity, and financial constraints, available treatment and disposal options are often not well designed, well built, or well managed.** According to the Stockholm Convention, there is a risk of PCDD [polychlorinated dibenzodioxins] and PCDF [polychlorinated dibenzofurans] being released in relatively high amounts if healthcare waste is incinerated in conditions that do not meet best available techniques or best environmental practices.

- Small healthcare waste incinerators, such as single chamber, drum, and brick incinerators, are built to satisfy the need for public health safety when more advanced systems are not feasible to introduce and maintain. If the only other option is unregulated disposal, this entails a balance between environmental impacts and other longer-term or unintentional negative health effects from managed combustion, with an overarching need for urgent public health security.
- In many developing countries, these conditions exist, and small-scale incineration may be a temporary solution to a pressing need (WHO 2014). To stop the production of dioxins and furans, burning PVC plastics and other chlorinated waste should be avoided as far as possible.
- Burning healthcare waste in a pit is less ideal, but it should be done in a small environment if it is really the only viable choice in an emergency or if chosen as an interim solution if no other option is available. The waste should then be burned in a dugout pit before being covered with soil (WHO 2014).
- Temporary disposal by onsite burial is often permitted due to the lack of a sanitary landfill or difficult access to one, while national and local municipal authorities must upgrade the current landfill or, if appropriate, create a sanitary landfill to ensure the safe disposal of waste in the region. Infectious waste should be stored in separate cells (e.g., a separate pit, a specially specified storage area, etc.) for the duration of the pandemic, covered with locally available material and with limited access.

16   Occupational Safety and Health

During the COVID-19 pandemic, healthcare waste handlers are the most vulnerable. Sharps that are not disposed of into puncture resistant containers put staff at risk of infection and injury. The risk of contracting a secondary infection after a needle-stick injury caused by a contaminated sharp is determined by the amount of contamination and the type of infection from the source patient. As a result, most countries have implemented stringent guidelines for Occupational Safety and Health (OSH) activities, which are based on national and international standards. The following are some of the findings from the desk report, as well as the results of the questionnaire survey.

- Provide required personal protective equipment, such as masks (3-layer masks, N95 masks, surgical masks), gloves (heavy-duty gloves), rubber boots, disposable work wear/hazmat suit, goggles, face shields, and hair cover/caps, to healthcare waste management staff, as well as advise and educate them on how to use it for personal safety.

- Inform and educate healthcare staff on good hygiene and safety procedures, such as using hand sanitizer/cleaning agents in collection vehicles, hand washing/cleaning before and after picking up waste, and disinfection (waste container/bags and waste collection vehicle).

17 Capacity Building and Awareness Raising

During the COVID-19 epidemic/pandemic, the most important point of view is "continuity," which means managing waste created by healthcare facilities, households, and quarantine locations while considering COVID-19 transmission risks in any waste management flow, even if the amount of waste is dramatically increased. COVID-19 waste can be treated similarly to a disaster response, according to the Japanese government's notification to local governments.

In this context, all stakeholders, including national and local governments, healthcare/laboratory facilities, and the public, must strengthen their capacity and awareness for COVID-19 waste management in emergency response, including contingency plan development and preparedness and build back better (BBB) for further improvement of healthcare waste management and MSW management. Training and capacity building can be done in the following ways:

Introduce training and education programs for workers of healthcare facilities on potential hazards from waste, the purpose of immunization, safe waste-handling procedures, reporting of exposures and injuries, use of PPE, and hygiene practice.

- **Training and education for logistics staff can be focused on the knowledge of risks and handling of hazardous waste**, including relevant legal regulations, waste classifications and risks, safe handling of hazardous waste, labeling and documentation, and emergency and spillage procedures
- **Training and education also need to focus on the drivers and waste handlers who are transporting healthcare waste** to the central treatment and disposal sites, informing them of the risks and handling of driving trucks with dangerous waste; in addition, verification indicating the confidence to transport hazardous waste is preferred.
- **In the time of the COVID-19 pandemic**, in addition to trainings on safe healthcare waste management, awareness raising on safe and healthy working environment for healthcare waste workers, including both for the formal and informal sectors, is required. For example:

  - Sick workers should stay home and practice proper sneezing and coughing etiquette, as well as proper hand hygiene; (ii) Regular workplace environmental cleaning; (iii) Healthy employees notifying managers if a family member is sick; and (iv) Employers notifying other employees if an employee is known to have COVID-19, for potential exposure.

- All employees must be educated on the dangers of virus exposure, the hazards associated with that exposure, and the proper workplace procedures to avoid or reduce the risk of infection.
- Job shifts should be updated, and strategies to minimize human contact and ensure distance between workers at work could be implemented.
- Consider collaborating with local waste pickers groups to buy personal protective equipment (PPE: gowns, gloves, masks), hygiene kits, and food supplies to help support the livelihood loss of informal waste staff.
- Awareness raising and communications for healthcare staff and the public are also required.
- Additional guidelines for the handling, care, and disposal of waste created during COVID-19 patient treatment/diagnosis/quarantine (public communications).
- Development of media (such as a Web site or a public service announcement) for waste management hygiene and safety.
- The new collection schedule and other relevant improvements in the MSWM system must be communicated through radio, newspapers, and other approved media. Citizens should dispose of waste according to the municipality's guidelines.
- To prevent the development of unregulated dumpsites and to allow emergency teams to effectively maintain sanitation in the region, disposal must take place only in designated areas.
- Restrictions and public awareness measures, as well as the distribution of personal protective equipment (PPE) to waste pickers working at the disposal site, should all be considered.
- Relevant stakeholders should be consulted on waste collection and treatment methods and recommendations in order to enhance collaboration, service efficiency, and the response plan.

17.1 Engage with All Stakeholders in the Society

- Waste management should be classified as a critical operation.
- Define positions and obligations in the organized work to ensure safe waste collection, care, and disposal by interacting and consulting with stakeholders in the waste stream, both formal and informal.
- Maintain and develop collection systems in low-income areas in partnership with informal staff, and use time to improve the informal sector's network.
- Staff shortage has to be accounted for and collection and treatment partners need to be included in assigning responsibilities.

18 Policy, Regulatory and Institutional Framework

- International guidelines and regulations for safe HCWM are available, and most countries have referred to and followed them. This lays a solid foundation for managing healthcare waste during the COVID-19 pandemic, especially waste created by healthcare facilities (whether existing or additional emergency healthcare facilities recently built).
- Additional regulations and guidelines for healthcare waste created by non-healthcare facilities, such as households and public spaces, are required, particularly in light of the growing generation of potentially polluted wastes (such as masks, tissues, and disposable clothes).
- During an emergency, such as the COVID-19 pandemic, a strong institutional structure is needed. For the collection of emerging COVID-19 waste from not only healthcare facilities, but also households and public places, clear roles and responsibilities must be allocated.
- COVID-19 waste management in emergency response, including contingency planning and preparedness, as well as build back better (BBB) for future progress Established policy and regulatory mechanisms must be reviewed and enforced to improve HCWM and MSWM. It is critical to conduct a rapid evaluation of the existing framework in order to recognize available capacities and weaknesses in the respective country or city, as well as to try to maximize the use of available treatment technologies to full extent [8].

19 Safe Handling of Infectious Waste

Although it is critical to anticipate the emergence of "special waste" produced during the COVID-19 pandemic, scientific findings such as the virus's viability on various materials (up to 72 h on plastics, 48 h on stainless steel, 24 h on cardboard, and 4 h on copper) should be taken into account in order to adjust existing waste management and thus proper waste handling.

- Proper segregation, packaging, and storage of potentially contaminated materials (double bag).
- Adjustment of collection frequency based on priority (organic waste, infectious waste, etc.) and possible reduction of collection of recyclables.
- Proper use of personal protective equipment (PPE) when handling healthcare waste, and hand hygiene as well as other precautionary practices to ensure health and safety of waste workers.
- Encourage all waste workers including formal municipality employees to comply using PPE.

Furthermore, from the standpoint of sustainability, special attention should be given in particular to the informal sector (which plays an important role in waste management in normal times), such as waste management continuity by mitigating infection transmission threats, social safeguards, OSH, and insurance, and so on.

20 Appropriate Treatment and Disposal Methods

Many factors must be considered when choosing treatment options, including national and international legislation, environmental and occupational safety, waste profile (characteristics and quantity), technology capabilities and specifications, costs, and operating and maintenance requirements. Although non-incineration waste disposal solutions should be used wherever possible, WHO urges all stakeholders to uphold the Stockholm Convention and work toward incrementally enhancing healthy healthcare waste management practices in order to protect human health and the environment (WHO 2007).

In fact, many municipalities and healthcare facilities in developing countries lack the most preferable technology for treating healthcare waste before disposal, and although using a landfill to protect public health is a viable alternative, the landfill is typically an open dumpsite with open burning and little operation and management with informal sector involvement. In this situation, intermediate or temporary treatment methods such as single chamber incineration without flue gas and automatic pressure pulsing gravity autoclaves are recommended at the very least. Although such technologies are not deemed suitable under international agreements since they do not meet their criteria, they can be recommended as a temporary solution. Developing countries with limited resources and considering an emergency situation, such as the COVID-19 pandemic, should use these temporary care options, which are listed below, in accordance with WHO Guidelines16:

1. No use of open dumpsites for HCW
   HCW should not be dumped on or near unregulated dumps. The danger of infectious diseases or dangerous materials coming into touch with humans, such as waste pickers and livestock, is evident, with additional risks of disease transmission by direct contact, wounds, inhalation or ingestion, as well as indirectly through the food chain or a pathogenic host species.
2. Minimum approach to treatment and disposal

   - This means, at the very least, segregation and other practices to reduce the amount of waste that must be handled.
   - A treatment method that achieves at least the necessary degree of disinfection, as well as proper disposal. Treatment waste, with the exception of sharps waste, may be disposed of with normal municipal solid waste (Table 7).

Table 7 A ladder of healthcare waste treatment technologies

Technologies in accordance with international conventions	Low—heat-based and chemical processes
	Dual chamber incineration with flue gas treatment
Interim treatment technologies	Dual chamber incineration without flue gas treatment
	Single chamber incineration without flue gar treatment
	Automated pressure pulsing gravity autoclaving
Uncontrolled waste combustion	Burning in a pit
	Open burning

Source Hygiene and Environmental Health Module (https://www.nih.gov)

- Hazardous healthcare waste from small healthcare facilities may be buried inside the grounds of the facility where public access can be limited and the burial site is well built-in severe situations where treatment is not necessary. It is recommended that a safe burial pit concept be used. Larger healthcare facilities can plan for a special cell or pit, regular soil cover, and limited access with a nearby landfill.

3. Use a site operated in a controlled manner that may already exist for MSWM
   Where a municipal waste landfill is available, it is possible to deposit HCW safely in two ways:

   - Immediately in front of the base of the working face where waste is being tipped, in a shallow hollow excavated in mature urban waste (preferably over three months old). When a load of HCW is deposited, it is filled by the advancing tipping face of fresh urban waste the same day (preferably creating a layer of municipal waste around 2 m thick). Picking up trash in this area of the site must be avoided. The same approach is commonly used for hazardous solid industrial wastes, with the intention of stopping animals and scavengers from re-excavating the waste after it has been deposited.
   - In a deeper (1–2 m) pit excavated in a mature urban waste covered field (i.e., waste at least three months old). After that, the pit is backfilled with the mature urban waste that was previously removed, as well as an intermediate soil cover (roughly 30 cm) or topsoil cover (up to 1 m). Picking up trash in this area of the site must be avoided.

4. Safe burial on hospital premises

   - In remote healthcare facilities and underdeveloped areas, minimal approaches to HCWM are needed. In addition, in temporary refugee encampments and areas experiencing extreme poverty, minimal practices may be needed. As a result, safe waste burial on hospital grounds may be the only viable alternative available at the time. Even in these hard times, hospital administration should develop the following basic principles:

(i) The burial site should be managed as a landfill, with each layer of waste covered by a layer of soil to prevent odors and contact with the decomposing waste, and to deter rodents and insects.

(ii) The design and use of a burial pit is illustrated below.

(iii) Once the pit is constructed, the safe burial of waste in minimal circumstances depends critically on staff following sensible operational practices. This must be insisted upon, and the local healthcare manager must realize their responsibility for making an organized waste disposal system work properly.

5. As an emergency response for treatment and disposal, WHO also introduced the following options.

- Burial in dumps onsite "Dig a pit that is 12–12 m wide and 2–3 m deep. The pit's bottom should be at least 2 m above the groundwater level. Clay or permeable material should be used to line the pit's rim.
- To keep water out of the pit, create an earthen mound around the opening.
- To prevent unauthorized entry, build a fence around the field. Alternate layers of waste within the trap, then fill with 10 cm of soil (if it is not possible to layer with soil, alternate the waste layers with lime). Cover the waste with soil and permanently seal it with cement and embedded wire mesh when the pit is within 50 cm of the ground surface" [9].

21 Burial in Special Cells in Dumping Sites (if Available in the Affected Area)

When burying waste in landfill sites, waste-containment cells may be used. At least 10 m long, 3 m high, and 1–2 m deep, the cell should be. The cell's bottom should be at least 2 m above groundwater. Soil or a low-permeability material should be used to cover the cell's rim. To avoid access by people or animals, the waste in the cell should be covered immediately with 10-cm layers of soil (in diseases outbreaks, preferably spread lime on waste before covering with the soil). To reduce public exposure to bio-contaminated wastes, it is highly recommended that HCW be transported in a safe manner.

Low-cost double chamber incinerators "Double chamber incinerators can reach a temperature of about 800 °C with a residence time of more than one second in the second chamber to kill pathogens and break down some of the particulates in the outlet gases to kill pathogens". The incinerators should be placed at a safe distance from structures. Before adding infectious wastes, such incinerators must be preheated with paper, wood, or dry non-toxic waste (small amounts of kerosene can be added if available).

21.1 Open Burning (WHO)

Burning medical waste (infectious waste) is less ideal, but if it is really the only viable choice in an emergency, it should be done in a small area (such as a dugout pit), and covered with a layer of dirt.

21.2 Temporary Storage (ADB)

- Safe facilities may be used as a transitional measure when waiting for additional emergency services to become available in the long term. 4.4 Capacity building and awareness rising.
- While most healthcare facility staff is qualified to manage healthcare waste, additional capacity building and awareness rising is desperately needed for families, the informal sector, such as waste pickers, and emergency healthcare facilities that use volunteers and temporary workers.
- Volunteer capacity building and user-friendly outreach materials, such as public service announcements on the radio and television/Web sites with regular presentations, can be used to raise public awareness about proper healthcare waste disposal.

22 General Principles and Guidance for Managing Infectious Waste During the COVID-19 Pandemic

Many countries and international development organizations have released general principles and detailed recommendations to handle healthcare waste during the COVID-19 pandemic, according to the desk report and questionnaire responses.

22.1 Government Instruction Regarding Waste Management During COVID-19

1. Ensure collection and treatment of healthcare waste.
2. Ensure the collection and treatment of household waste.
3. Maintain in operation incineration and land filling for non-hazardous waste.
4. Maintain in operation hazardous waste treatment.
5. Maintain as long as possible the separate collection of household waste (packaging, paper, cardboard, glass).

6. Maintain as long as possible in operation the sorting facilities for separately collected household waste.

23 Safe Management of Healthcare Waste During COVID-19

Best practices for efficiently handling healthcare waste should be followed, including delegating liability and allocating adequate human and material resources to properly segregate and dispose of waste. There is no indication that the COVID-19 virus has been transmitted by direct, unprotected human contact during the processing of medical waste. All healthcare waste produced during patient care is considered infectious (infectious, sharps, and pathological waste) and should be collected safely in clearly marked lined containers and sharp secure boxes, including those with documented COVID-19 infection. This waste should be treated, preferably onsite and then safely disposed. If waste is moved offsite, it is critical to understand where and how it will be treated and disposed.

Non-hazardous waste produced in waiting areas of healthcare facilities should be put in strong black bags and sealed completely before being collected and disposed of by municipal waste services. All who treats medical waste should put on the necessary PPE (boots, long-sleeved gown, heavy-duty gloves, mask, and goggles or a face shield) and wash their hands afterward. During the COVID 19 epidemic, the amount of infectious waste is expected to rise, owing to the widespread use of personal protective equipment (PPE).

This involves pumping out tanks and unloading pumper trucks for crews. Individuals should safely remove their PPE and practice hand hygiene before entering the transport vehicle after handling the waste and until there is no chance of further exposure. Soiled PPE should be put in a sealed bag to be laundered safely later. In the absence of offsite treatment, in-situ lime treatment may be used. Ten percent lime slurry is applied at a rate of one part lime slurry per ten parts waste in this treatment.

24 Safe Management of Dead Bodies During COVID-19

Although the risk of COVID-19 transmission from handling a deceased person's body is minimal, healthcare workers and those who come into contact with dead bodies should take standard precautions at all times. Scrubs, impermeable disposable gown (or disposable gown with impermeable apron), gloves, mask, face shield (preferably) or goggles, and boots should be worn by healthcare personnel or mortuary staff who are preparing the corpse. PPE should be properly discarded after use and decontaminated or disposed of as infectious waste as soon as possible, as well as hand hygiene. The body of a deceased person confirmed or suspected to have COVID-19 should be wrapped in cloth or fabric and transferred as soon as possible to the mortuary area.

25 Guidance on Management of Household Waste in COVID-19 Cases

An individual waste bag should be placed in the patient's room.

- Paper tissues and face masks used by the patient should be immediately put in the waste bag that was placed in the patient's room.
- Gloves and face masks used by the caretaker and by the cleaner should be immediately put in a second waste bag, placed near the door to the patient's room, when the caretaker or cleaner leave.
- Before being removed from the patient's room, the waste bags should be closed and replaced frequently; they should never be emptied into another container.
- The open patient waste bags can be gathered and placed in a clean general garbage bag; the closed patient waste bags can be thrown away in the unsorted garbage. There is no need for a special collection operation or other form of disposal.
- After handling waste bags, strict hand hygiene should be performed: use water and soap or alcohol-based hand disinfectants.

26 COVID-19: Revised Guidelines Show How Biomedical Waste Must Be Handled

- Concerns about biomedical waste created by treating novel coronavirus disease (COVID-19) patients, sanitation workers who handle this waste, and extra care to be taken by isolation wards were emphasized in revised guidelines released on June 10, 2020, by the Central Pollution Control Board (CPCB).
- This was the third time the guidelines have been updated since they were first released on April 18. The new guidelines were developed based on existing expertise and practices in the management of other infectious waste produced in hospitals during the treatment of viral and contagious diseases.
- All stakeholders, including isolation wards, quarantine centers, sample collection centers, labs, Urban Local Bodies (ULBs), and general biomedical waste treatment and disposal facilities, must meet these guidelines (CBWTFs).
- The guidelines add to existing practices under the Biomedical Waste Management Rules, 2016.

- The revision incorporates guidance on segregation of general solid waste and biomedical waste generated by ULBs, CBWTFs, quarantine centers and healthcare facilities treating COVID-19 patients.

27 Safety of Sanitation Workers

- The safety of waste handlers and sanitation workers associated with such healthcare facilities.
- The guidelines emphasize extra care to be taken at COVID-19 isolation wards. Foot-operated lids in color-coded bins must be introduced to avoid contact, according to the guidelines.
- General solid waste like medicine wrappers and cartons, syringes, fruit peels, empty bottles, discarded paper and other items not contaminated by patients' secretions and body fluids must be collected separately, according to Solid Waste Management Rules, 2016.
- Wet and dry solid waste bags must be securely tied and handed over to waste collectors authorized by ULBs daily.
- Non-disposable items must not be disposed of as much as possible and should, instead, be cleaned and disinfected keeping hospital rules in mind, the guidelines said.

28 Waste Segregation

Leftover food, discarded dishes, glasses, used masks, towels, toiletries, and other items used by COVID-19 patients are categorized as biomedical waste and should be disposed of in yellow bags, whereas used gloves should be disposed of in red bags. "This distinguishes between COVID-19 waste that must be incinerated and COVID-19 waste that can be disinfected, autoclaved (a method that destroys bacteria, viruses, and other microorganisms), and disposed of." This helps to minimize the amount of COVID-19 waste produced while also reducing the burden on CBWTFs for incineration. According to the guidelines, appointed nodal officers for biomedical waste management in hospitals must be responsible for training waste handlers on infection prevention measures.

- Hand hygiene, respiratory etiquette, social distancing, and the use of acceptable personal protective equipment are among the steps that must be illustrated by videos and in local languages.

- Health departments and technical organizations, in collaboration with state pollution control boards (SPCBs) or pollution control committees, must train nodal officers (PCCs).
- People who operate quarantine camps, families, or homecare facilities are responsible for handing over general municipal solid waste to waste collectors listed by ULBs.
- Waste generated from kitchens, packing material for groceries, food material, waste papers, waste plastic, floor cleaning dust, etc., handled by patient care-takers or suspected quarantined individuals should be treated as general waste, the guidelines said.
- Waste contaminated with blood or body fluids of COVID-19 patients must be collected in yellow bags. The storage of general waste in yellow bags is strictly not allowed.

29 Role of Nodal Officers

- The guidelines authorize ULBs to employ specialist waste management firms to collect solid and biomedical waste in a timely manner if collection and transportation operations are insufficient due to current staffing levels.
- Through the CPCB's COVID19BWM biomedical waste-tracking mobile application, all quarantine centers were required to quantify and monitor the movement of COVID-19.
- Every day, nodal officers of quarantine centers must update the regular generation of COVID-19 waste.
- "Prior to the revised guidelines, we received all kinds of mixed waste in a yellow bag from containment zones and quarantine households," said a CBWTF operator in Gurgaon. The operator added, "The bottom ash content from the mixed waste after incineration is very high."
- According to the operator, the revisions in the guidelines seek to reduce the burden on CBWTFs while maintaining the safe disposal of COVID-19 waste [10].

30 Conclusion

As a result, increasing capacity to manage and treat this healthcare waste is critical. Additional waste treatment capability, ideally through alternative treatment technologies like autoclaving or high-temperature burn incinerators, may be required, as well as systems to ensure their long-term operation. If latrines and holding tanks of excreta from suspected or confirmed COVID-19 cases are complete, there is no need to empty them. Best practices for properly handling excreta should be practiced in general. Latrines or holding tanks should be built to meet patient demand, taking into account the likelihood of unexpected increases in cases, and they should be drained on a regular basis based on the wastewater volumes produced. When treating or transporting excreta offsite, PPE (long-sleeved robe, gloves, boots, masks, and goggles or a face shield) should be worn at all times, and great care should be taken to prevent splashing.

References

1. Charan J, Kaur R, Bhardwaj P, Kanchan T, Mitra P, Yadav D, Sharma P, Misra S (2020) Snapshot of COVID-19 related clinical trials in India. Indian J Clin Biochem 35:418–422
2. Pak A, Adegboye OA, Adekunle AI, Rahman KM, McBryde ES, Eisen DP (2020) Economic consequences of the COVID-19 outbreak: the need for epidemic preparedness. Front Public Health 8:241
3. Mullick AR, Bari S, Islam MT (2020) Pandemic COVID-19 and biomedical waste handling: a review study. JMSCR 08(05):497–502
4. Das A, Garg R, Banerjee T (2020) Biomedical waste management: the challenge amidst COVID-19 pandemic. J Lab Phys 12(2):161–162
5. van Dorn A (2020) COVID-19 and readjusting clinical trials. World Report 396(10250): 523–524. https://doi.org/10.1016/SO140-6736120-31787-6
6. Singh H, Rehman R, Bumb SS (2014) Management of biomedical waste: a review. Int J Dental Med Res 1(1):14–20
7. Tsukiji M, Solihin I, Pratomo Y, Onogawa K, Alverson K, Honda S, Ternald D, Dilley M, Fujioka J, Condrorini D (2020) Waste management during the COVID-19 pandemic from response to recovery financial support. This report was developed with financial support from the Ministry of Environment (MOE), Government of Japan through United Nations Environment Programme—International Environmental Technology Centre (UNEP-IETC)
8. Bandela DR (2020) COVID-19 revised guidelines show how biomedical waste must be handled. Reported by Municipal Solid Waste Management, Centre for Science and Environment, New Delhi
9. Ramesh Babu B, Parande AK, Rajalakshmi R, Suriyakala P, Volga M (2009) Management of biomedical waste in india and other countries—a review. Int J Environ Appl Sci 4(1):65–78
10. Capoor MR, Bhowmik KT (2017) Current perspectives on biomedical waste management: rules, conventions and treatment technologies. Indian J Microbiol 35(2):157–164

Websites

11. https://www.ncbi.nlm.nih.gov/pmc/articles/PMC7467828/
12. https://swachhindia.ndtv.com/india
13. http://www.ijmedrev.com/article_120913.html
14. https://cpcb.nic.in/covid-waste-management/
15. https://www.investindia.gov.in/team-india-blogs/bio-medical-waste-management-during-covid-19
16. https://www.who.int/docs/default-source/searo/
17. https://pib.gov.in/PressReleasePage.aspx?PRID=1694438

18. https://www.moneycontrol.com/news/opinion/covid-19-the-growing-challenge-of-biomedical-waste-disposal-6492401.html
19. https://www.businesstoday.in/current/economy-politics/-covid-19-biomedical-waste
20. https://covid.aiims.edu/biomedical-waste-management-disinfection-cleaning-in-covid-19-areas/

Personal Protective Equipment for COVID-19

M. Gopalakrishnan, V. Punitha, R. Prema, S. Niveathitha, and D. Saravanan

Abstract COVID-19 has resulted a change in the mindset of human beings and a strong emphasis has been spread among the public that precaution is the better way to protect oneself from this pandemic. Various personal protective equipments (PPE) are used to protect human beings from COVID-19. Requirements of PPE for healthcare professionals differ from that of COVID affected victims and healthy persons. Standards for the care of COVID affected patients are recommended by the World Health Organization (WHO). Protection from physical contacts and airborne salivary droplets are essential to protect from infectious virus, which is largely met by commercially available masks with different efficiency levels. Personal protective equipment including surgical gloves, goggles, glasses, face shields, particulate respirators, powered respirators, single or multi-layered masks, ventilators, surgical suits, tops, aprons, scrubs, heavy-duty alcohol-based hand rubs/sanitisers, hand drying tissues, biohazard bags and fit-test kits are recommended by the WHO to safeguard the human being. A detailed study of the personal protective equipment for COVID-19 has been reported in this chapter.

Keywords Breathability · Filtration · Layered masks · Respirators · Surgical mask · Covid-19 · Virus · sanitiser · Hand wash

M. Gopalakrishnan (✉) · V. Punitha
Department of Textile Technology, Bannari Amman Institute of Technology, Sathyamangalam 638401, India

R. Prema
Department of Electrical and Electronics Engineering, Bannari Amman Institute of Technology, Sathyamangalam 638401, India

S. Niveathitha
Department of Fashion Technology, Bannari Amman Institute of Technology, Sathyamangalam 638401, India

D. Saravanan
Department of Textile Technology, Kumaraguru College of Technology, Coimbatore 641049, India

1   Introduction

A disease is declared as an epidemic when an outbreak of disease spreads over a wide area, affecting many individuals at the same time to become ill, while the epidemic can potentially become a pandemic when wider geographical areas and significant (or exceptionally high) proportion of the population are affected. Since, the emerging of covid-19 virus in china, the world health organization (WHO) declared a pandemic and survival becomes difficult [1]. To control the growth of covid-19 a severe acute respiratory syndrome corona virus (SARS-CoV-2), various precautions have been made by the government [2]. One such and most effective initiative is complete closure of living spaces and boundaries or widely known as "lockdown" (*stay-in-home*). Apart from the social distancing, wearing the right kind of face masks reduce the spread of COVID-19 [3–5]. The health care professionals use the face masks of disposable types to prevent occupational hazards. During, SARS and H1N1 outbreaks, the people used face masks with different levels of protection [6, 7]. In this current COVID-19 pandemic conditions, the use of face mask may help in reducing the spread of virus until the proper mode of protection for COVID-19 understood [8–11].

2   Viruses—Strong and Mutative

In biology, viruses are regarded as non-living, small infectious agents, which invade living cells (host), multiply inside those cells and causing damages (illness/ diseases) to the hosts, e.g. animals, plants, humans, bacteria and cause diseases [12]. The viruses are either intracellular (active form) or extracellular (inactive form), adapted to transfer the nucleic acid from one cell to another (Fig. 1). Classification of viruses are based on their (i) structure (simple—made up of nucleic acid and protein shell, complex—made up of nucleic acid, protein shell and lipoprotein envelope) and (ii) type of nucleic acid (DNA or RNA). The nucleic acid can be single or double-stranded, protected by a shell containing proteins, lipids, carbohydrates, or their combinations [13, 14]. Occurrences of single-stranded DNA viruses are rare, while twofold stranded DNA viruses are commonly observed whereas in the case of RNA viruses, there are very few cases of twofold stranded RNA viruses and predominantly they are single-stranded.

A virus attaches to a living cell and injects its nucleic acid into it where the nucleic acid binds to the ribosomes of the cell and stimulates them to produce viral proteins [12, 14]. Multiplication of the viral genome occurs through replication, resulting in a huge number of new copies of the viral RNA or DNA, i.e. new viruses. The host cells are damaged in this process and become no longer beneficial to the viruses, which makes the viruses leave those cells and target new cells.

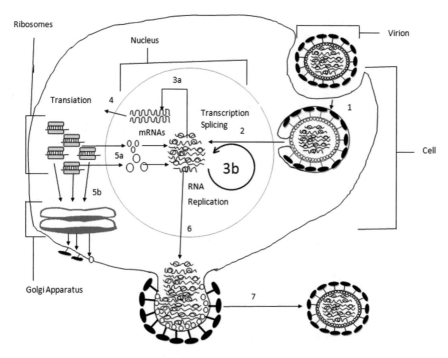

Fig. 1 Action of virus in host cell

3 DNA Viruses Versus RNA Viruses

Deoxyribonucleic acid (DNA), double-stranded molecules found in the nucleus, is the major storage for genetic codes that contains information for the functioning and advancement of all living organisms. DNA virus instils the genetic code specifically to the membrane of the host DNA then with the help of RNA polymerase, the duplication happens in the nucleus and released during the lytic phase of the host cells (multiplication step) with the copies of infection. Since the specificity of the DNA viruses are detected at the transcriptional (constant) level, certain vaccines are effective throughout the years. Ribonucleic acid (RNA) that contains ribose sugar, is usually a single-stranded molecule, instilling the RNA to the host cell cytoplasm. Unlike DNA viruses, which must always transcribe viral DNA into RNA to synthesize proteins, RNA can skip the transcription process [15]. DNA, here, acts as a pattern for RNA and transcribes it into viral proteins. Certain RNA viruses embed transcriptase enzyme that transfer RNA virus to DNA virus and combine with the host DNA thereby following the DNA replication process. Mutation is the major cause of the changes, by RNA polymerase, in the genetic code of the viruses and makes them unstable, replace the protein coat that can confuse the immune system [13, 15]. When RNA viruses attack the human

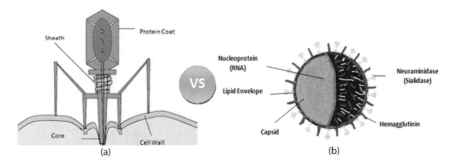

Fig. 2 Schematic representation of **a** DNA and **b** RNA viruses

living being, they infuse their RNA into the cytoplasm of the host cells, where RNA can be utilized to integrate proteins and frame the imitations (Fig. 2).

In DNA viruses, there are two stages in the translation procedure first, the mRNAs (messengers) are made (alpha and beta mRNA) subsequently, gamma mRNAs and are interpreted into the cytoplasm, further leading to DNA replication. These stages can't be recognized in the RNA interpretation process of RNA viruses, which interpret mRNAs on host ribosomes and make viral proteins instantly with higher transformation rates than DNA change rates [16]. This obviously results in faster communication of diseases in the case of those caused by RNA viruses.

Single-stranded RNA viruses can be further ordered into negative-sense and positive-sense RNA viruses, depending on the sense or polarity of the RNA. Positive (or plus-strand) and negative (or minus-strand or anti-sense) sense RNA viruses are the two types of single-stranded RNA viruses classified based on the type of genome [17]. Both positive and negative-sense RNA viruses are infectious and cause diseases in animals and plants. Positive-sense RNA viruses account for a large fraction of known viruses, including many pathogens such as hepatitis C virus, dengue virus and SARS and MERS coronaviruses, as well as less clinically less serious, common cold.

4 Coronaviruses

Coronaviruses are a group of viruses that can cause a range of symptoms including a running nose, cough, sore throat and fever, usually spread through direct contact with infected persons. Coronaviruses, name originating from Latin for crown-spikes on its surface, have helically symmetrical nucleocapsids (nucleic acid and surrounding protein coat), which is uncommon among positive-sense RNA viruses, but far more common for negative-sense RNA viruses (Fig. 3).

Human coronaviruses were first identified in the mid-1960s and there are four main sub-groups of coronaviruses, known as alpha, beta, gamma and delta. Sometimes coronaviruses that infect animals can evolve, make people sick and

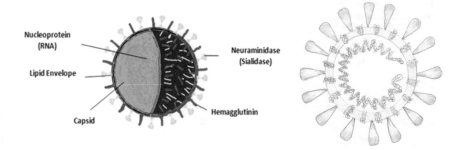

Fig. 3 Schematic representation of influenza and corona viruses

become a new human coronavirus like MERS-CoV (the beta coronavirus that causes Middle East Respiratory Syndrome), SARS-CoV (the beta coronavirus that causes severe acute respiratory syndrome) and SARS-CoV-2 (COronaVIrus Disease 2019, or COVID-19). Including the newly identified form of the virus, there are a total of seven coronaviruses that can infect humans [12, 18, 19]. All the coronaviruses can be transmitted between humans through close contact. MERS, which was transmitted from touching infected camels or consuming their meat or milk, was first reported in 2012 in Saudi Arabia. SARS was first reported in 2002 in Southern China was thought to have spread from bats that infected civets. Since the virus, nCov 2019, first popped up in Wuhan (China) among the people who had visited a local Huanan seafood market, it is hoped to have spread from animal to humans. In a recent study, the researchers compared the 2019-nCoV genetic sequence and found close relationship (88% of their genetic sequence) with the viruses that originated in bats.

Invariably, many of us get infected due to coronaviruses in certain stages with mild to moderate symptoms, respiratory tract illnesses including pneumonia and bronchitis. These viruses are common amongst animals worldwide and rarely, coronaviruses can evolve and spread from animals to humans, the typical cases include MERS and SARS-Cov. It is unclear how the new coronavirus compares in severity, as it causes severe symptoms and death in some patients while causing only mild illness in the case of others (Fig. 4).

Pandemic effect in humans, by the viruses, are caused by a cycle that involves (i) infection among human, (ii) replication among human and (iii) rapid spread among the human and right now, it is very much unclear how easily the virus spreads from person to person. The human coronaviruses are supposedly spread from an infected person to others through (i) the air—respiratory droplets when a sick person coughs or sneezes, (ii) close personal contact (touching or shaking hands), (iii) an object or surface with viral particles (subsequent human contacts through mouth, nose or eyes before washing your hands) and (iv) rarely from fecal contamination [18]. These virus particles are pushed into the atmosphere and eventually affect air quality [21]. Poor ventilation, dirty air conditioner or HVAC system help speed up the spread of bacteria and viruses, in general [22].

Fig. 4 Impact of viruses in terms of spread and effect [20]

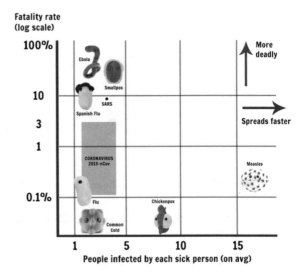

COVID-19 is said to spread in a country or a space, following a four different stage namely (i) stage 1—the country is not the source of the pandemic, still reports first few cases of the disease due to virus import; stage 1 happens when the epicentre of virus outbreak unable to contain the virus spread; (ii) stage 2—happens when local transmission of the disease is reported, spreading through the people of a particular country itself but possible to identify the trajectory of the virus; (iii) stage 3—happens when community transmission of the disease is observed and difficult to track the chain of transmission and stage 4 signifies the pandemic reaching the epidemic stage, affecting the masses and the measures to contain the virus and control the affected persons become very difficult.

5 Face Mask

Masks are often used by healthcare providers in hospitals and clinics, where the risk of transmitting or catching an illness is high. In healthcare setup, primarily masks are used to isolate from patients as well as protect the patients from potentially spreading the infections. There are two types of masks used in preventing infections: surgical masks and respirator masks.

Face mask is one personal protective equipment widely used nowadays to protect ourselves from the airborne virus and liquid contaminants. The N95 respirator mask is not meant for general public and it is recommended for the front line doctors and health care operators, who is required to protect themselves from the respiratory diseases like corona virus. The various regulatory bodies regulates the limits of N95 respirator mask, including OSHA (Occupational safety and health administration), NIOSH (National institute for occupational safety and health) and

CDC (Centre for disease control and prevention). The regulatory bodies regulate the N95 respirator that should prevent airborne viruses optimally [23]. In N95 face mask, the 95 indicates the percentage of 0.3 and more micron particles prevented from the penetration. So it indicates the efficiency of the micro particles removed or protected. N99 indicates 99% efficiency in removing the 0.3 plus microns and some may be indicated as 100 which means 99.97% of 0.3 plus micron particles are removed [24].

This N95 mask prevent the dust, fumes and mist of particle size more than 0.3 microns as per the regulation of CDC. The 'N' indicates the rating of respirator and it indicates non-oil that means it can be used where the oils are not present. The other variant in N95 mask is resistant to oil (R variant). This can be used where the oil-based particles are present. This is effective for a period of 8 h. 'P' is another variant of this class and the 'P' indicates the oil proof. The polypropylene non-woven fabric, produced from electrostatic method has better filtration efficiency, is used in N95 face masks. Exhalation valve is provided in certain N95 face masks and these valves increasing breathing by reducing the exhalation resistance [25].

Surgical masks give protection for nose and mouth and are mostly disposable in nature, thereby giving freedom in fitting the mask with face.

As per the Health ministry, wearing of face mask is compulsory for public in a pandemic situation like COVID. The masks may be made of textile materials possibly, cotton which fastened or cover the mouth and nose. WHO has revised the guidelines for wearing the face mask for public in June 2020 and people, in the public area, are insisted to wear the face mask made up either cloth or surgical mask for public and medical mask if the person is affected by COVID-19 [26].

Pathogens have the power to transmit through air and this was first identified by Wells [27–29]. The droplets of contagious particles are spread from a sneeze or cough when a person has an infection [30, 31]. Until the SARS in 2003 and swine influenza in 2009, the airborne transmission was not taken seriously. Various studies were taken, after the pandemic in 2003 and 2009, to analyze the mechanism to reduce or control the growth of the airborne virus. As per the WHO in 2014, the droplet's diameter of sizes greater than 5–10 micro metre normally called respiratory droplets and the diameter smaller than 5 micro metre is called droplet nuclei, is the cause for respiratory infections. The microbes will remain for a longer time and its transmitted distance will be more than one metre for droplet nuclei than droplet transmission [26]. Airborne transmission occupies the important role in propagating the spread of the virus as like in SARS.

6 Bamboo Nonwoven Based Face Mask

Healthcare textiles receives huge attention in this COVID-19 pandemic situations [32, 33]. The protection against microbes, people uses sanitisers and even when they wash their cloths, certain antimicrobial agents are used in the rinsing stage [34]. This is mainly to improve the protection against the pathogenic microbes.

Various fibres and fabrics were used in antimicrobial textiles. Some fibres are very poor against the microbial attack [35] whereas, some fibres are very good [36] against the pathogenic microbes. Cotton is the fibre which is mainly used in apparels [37] possess poor antimicrobial activity under favourable conditions like, temperature, humidity and warmth that play vital roles in the growth of the microbes [36–40]. Bamboo is a fast-growing plant with cellulose content of 30–40% [41], which provides excellent protection against the pathogenic microbes [42, 43]. The bamboo fibres are not only good in protection against the pathogenic microbes, but also shows good wicking characteristics [44], UV protection and soft handle, often made from spunlace [45]. Many methods for the production of bamboo fabrics are available, but the most common and widely used method is 'bamboo Viscose', a method similar to viscose manufacturing [46]. The spunlace fabric is used in the production of wipes which are antibacterial in nature with good absorbency. Though the bamboo fibre has antibacterial properties by nature during processing it losses its antibacterial nature. Key properties of bamboo fibre include (i) smooth and luxurious to touch, (ii) good breathability, (iii) cool and comfortable to wear, (iv) soft and drape, (v) strong and durable, (vi) antistatic and (vii) abrasion-resistant. Similarly, chitosan has good antimicrobial properties, finds the applications in food science due to its bio-degradability, non-toxicity, bio-compatibility [47, 48]. The face mask produced with bamboo fabrics are coated with chitosan due to their antimicrobial and the moisture vapour permeability activities of the chitosan and bamboo [48]. On treating the fabrics with chitosan makes the fabric resistant to microbial attacks. The term antimicrobial refers to the antibacterial and antifungal and not antiviral properties. The dissolved chitosan is applied on bamboo fabrics by padding along with a binder. This method not only fixes the chitosan on the surface of the fabric but also increases the properties like wrinkle resistance and anti-odour. The colloidal solution for coating is prepared by mixing chitosan and $AgCl-TiO_2$ with 1:5 ratio. The prepared bamboo fabric is immersed into the chitosan-based colloidal solution and squeezed to have 70% of the pick-up. The squeezed fabric is dried below 120 °C for 2 min and cured at 150 °C for 3 min. AATCC Test Method 100 (Agar diffusion method) using *Staphylococcus aureus* and *Escherichia coli* and ASTM F2299 method—Particle Filtration Efficiency (PFE) exhibit good in antimicrobial activity and particle filtration efficiency [49, 50].

7 Ventilators

In the treatment against contagious COVID-19 disease, ventilators are the special life-saving equipment availed all over the world. Lungs are the more susceptible part of the human body for the COVID-19 virus that infects the lungs at walls and air sacs regions. It has been observed that on 5% of affected patients, virus damages were found in the lungs wall and linings of air sacks, leading to reduced breathing ability which in turn result in lack of oxygen supply to the blood cells. In this critical illness, one needs to assist with external supportive care to enhance respiratory function to maintain the adequate oxygen supply to the blood cells [51]. The physician checks and monitors the blood oxygen level using a pulse oximeter which is clipped to the finger of the patient and then decides to provide mechanical ventilation if the oxygen level declines continuously for a certain time.

Ventilators are devices that serve the purpose of artificial ventilation by blowing oxygen into the lung of the patient undergoing inhalation difficulty (Fig. 5). This device includes input and output pipes for oxygen supply and a plastic tube called endotracheal tube which can be placed through the mouth into our respiratory tube [52, 53]. The additive manufacturing society has come up with 3D printing prototype machines to produce the respiratory valve to connect the ventilators and respiratory tubes. Three Briggs T-tubes can be joined to produce a valve connecting four sets of ventilator or respiratory tubes. This enables the effective utilization of the available ventilators for different patients simultaneously [54]. The Food and Drugs Administration (FDA) issued emergency use authorization for certain additional ventilators, converting modified anesthesia gas machines and modified positive pressure breathing devices to be used as ventilators on meeting the specified safety, performance and labelling criteria [55]. The FDA also released criteria for safety, performance and labelling information for emergency utilization of ventilators, conformity standards for ventilator devices, device specifications and usage instructions for ventilators and accessories and reprocessing and self-life information. FDA manual also provides guidelines for multiplexing adapters for splitting the ventilator support for more than one patient [56].

The primary task of ventilators is accomplished by three functions such as removing heat from the inlet air as the environmental air present at high room

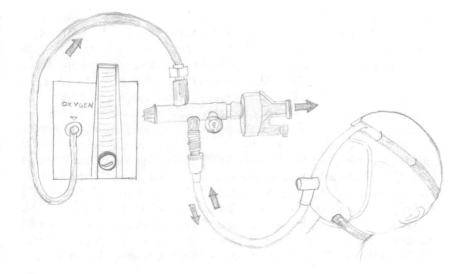

Fig. 5 Schematic representation of a ventilator

temperature, removing contaminants from the input air moreover supplying fresh air to the respiratory zone. The ventilators protect the occupant from other cross-infection from the environment [57]. Engineers of MIT redesigned a device [58] based on manually operating bag-valve-mask for manual bagging when all the ventilators in a hospital focussed on severely suffering patients and still a new patient need to be assisted with respiratory support. This device is an open-source project and it is not under the scope of FDA as this lacks monitoring the respiratory oxygen level and intake level of the patient [58].

The ventilators must focus on the operating parameters which are set to be in control to effectively support the respiratory function with the focus of life-saving. To ensure the efficiency of the ventilators in the life-saving process, the following operating factors need to be closely maintained accurately [59], (i) respiratory rate (RR), (ii) tidal volume, (iii) ratio of inspiratory (I) to expiratory (E), (iv) Positive end-expiratory pressure (PEEP). Acute Respiratory Distress Syndrome (ARDS) is ventilation-induced lung damage caused by mechanical ventilators. This ventilator-induced respiratory damage occurs at both high and low tidal volume of ventilation. The high tidal volume of oxygen supply leads to alveolar tissue rupture and is followed by air leaks at high pressure due to over-distention. Low tidal volume ventilation induces lung damage also called atelectrauma caused by improper opening and closing of parts of the lung and the injury is inflated by a higher level of inhomogeneity [60]. In the effort to reduce or avoid the ARDS in a patient, a ventilator should be set to support respiration with less tidal volume, the prone position of ventilation, High PEEP and recruitment manoeuvres. And, the ventilators having features of electronic monitors and sensors, automatic alarms for safety measures along with training information and user manuals play a major role in reducing these issues [61].

Spiro wave is another design followed by MIT's manual bagging model, an automatic resuscitator developed with basic pressure control sensors and alarms, controlled tidal volume and other features instructed by FDA in EUA. University of Minnesota's developed a device called *coventor*, which is comparable with bag-valve type ventilators but operates without an external oxygen supply. This is small in size and comparatively, the low cost can be made of metal stamped or 3D moulded frame compiles on FDA's Emergency use authorization guidelines [62].

NASA Jet Propulsion Laboratory compiled functional features of respiratory support with the inputs had from discussion among respiratory expertise physicians and therapists. The features focussed on the critical operating conditions such as controlling the Peak Inspiratory Pressure, PEEP (Positive Expiratory End Pressure), achieving maximum volumetric flow rate, operating with controlled breathing rate (BPM) and minimum operating hours without the intervention of at least 24 h. They invented Ventilator, Intervention, Technology Accessible Locally (VITAL) device with a key feature of non-resterilization compiled on FDA's emergency use authorization critical usage instructions and guidelines for utilization during COVID-19 pandemic. This device is made to operate at a controlled pressure, tidal volume, inspiratory to expiratory ratio, PEEP and RR with specifically designed VITAL software [59].

8   Surgical Suits

SARS—COVID coronavirus is a highly contagious pathogen that can reach 8 m of distance from the patient through aerosol when sneezed out. The droplets of size range from 1 to 5 mm are generally discharged by people while sneezing or coughing and it can hang ahead of 1–2 m in air space. In these exceptional times of treating COVID-19 viral disease patients, all doctors and clinicians ensure the surgical amenities to avoid contact with the virus from the affected person [63]. Operating rooms (OR) for imperative surgical treatments of COVID infected patients are exclusively dedicated in each institution. These rooms are enclosed with a protective system and the entrance and exit portions are air filtered using a HEPA filter. The particulate size of the COVID-19 virus is about 125 nm and the HEPA filter is 100% efficient to filter these particles from air circulation.

In this specific scenario, the hospital should divide the operative complex into five zones for ensuring the protection of physicians and staff. The first zone is the entry room where personal protective equipment is done followed by the second zone where disinfection and dressing of surgical suits are done. These dedicated zones helps in preventing the street contaminations from entering the OR. The third room is the operating room for exclusive COVID-19 patients. Exit rooms are divided into two sections where the fourth zone for removing PPEs and followed by the final zone where the staff can completely disinfect or shower themselves [64] (Fig. 6).

In the operation room, the patient or infected body acts as a source of threat and unless the healthy workers like doctors, clinical staff are entirely protected, they are also infected. Transmission-Based Precautionary measures are imperative to avoid transmission of airborne and droplet transferable pathogens, which can be achieved by personal protective suits and surgical suits. There are different ways of transmission of infectious pathogens takes place from affected person to healthy bodies inside hospital environment and other medical institutions. These are, direct and Indirect contact transmission, droplets (Sneezing or coughing, blood strike-through, etc.), airborne and common vehicle transmission and vector-borne transmission. Direct contact transmission involves shifting infection or microbes by touching the patients' infection source of wound without protective suits, whereas indirect contact transmission takes place by touching the intermediate contaminated equipment, gloves/clothes and objects or body parts. Common vehicle represents the equipment used to carry foods, water, medications and drugs, where contaminated vehicles act as transmitting agents or systems of hazardous pathogens. Vector-borne applies to flies, mosquitoes, rats and other microorganisms that can also transmit infections. But this vector-borne way of transmission is less significant in the hospital environment [66, 67]. The personal protective equipment and surgical suits are designed to prevent and control the infection spread as a combined enterprise of following suits in all the transmitting ways [68].

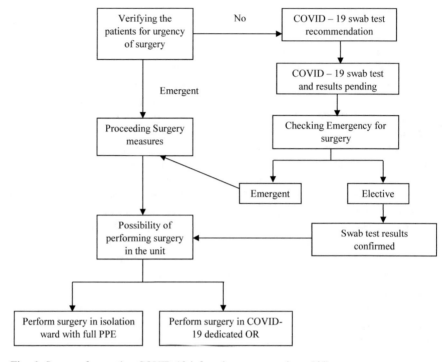

Fig. 6 Strategy for treating COVID-19 infected emergent patients [65]

- Surgical gowns and scrub
- Surgical aprons
- Eye protection goggles
- Face shields
- Respiratory mask
- Surgical and protective gloves
- Head protection cover
- Rubber footwear.

9 Surgical Gowns and Scrub

In all the emerging pandemic situations surgical gowns are the second-largest disposable suit used among medical physicians and health care workers followed by the surgical gloves. Previously reusable and greatly permeable loosely woven cotton and other conventional fibre clothing were used to protect the patient from surgical physicians and staff members. With increased concerns about the patho-gens that can transfer by blood strike-through, the purpose of the surgical gown has changed also to protecting surgeons and their team from patients. To proceed with the materialization of the surgical gown with protection from blood bear transmissible hazardous pathogens and wearer comfort for long-term operations it was proposed to focus on the pore size of fabric with the fabricated feature of non-permeability for liquids but breathable. The surgical gowns gained interest as the desire of physicians and health care workers in barrier consent to the spreading of peeled skin scales in the air atmosphere of operating room and avoids the transfer of infections through direct contact to the surgical team from patients wound and vice versa [69].

The fabric with a pore size lesser than the size of pathogens can arrest the penetration but producing a fabric of absolute pore size all over the material is a manufacturing challenge. Further, the surgical gown is characterized for certain properties to acquire impermeability against pathogens and wearer comfort. The general requirements (1–6) and fibre related properties (7–11) include the following,

1. resistance against liquid penetration (blood, serum, etc.,)
2. spotlessly cleanliness
3. flexibility/conformability
4. economical
5. breathable/air permeable
6. tactile softness
7. linting resistance
8. mechanical resistance
9. flame resistance
10. static safety and
11. toxicity.

In the absence of a complete surgical barrier gown, a foil gown can be used under the aprons. In addition to its barrier properties against infectious agents, the clothing material for protection against biological agents need to exhibit adequate mechanical resistance both in dry and wet conditions in terms of abrasion, tear, burst and puncture among other things. The material properties of surgical products are presented in the standard PN EN 14,126 as classes as in most European standards [70]. There are various combinations of fabrics that can be used for surgical gowns and categorized as reusable and disposable based on material type. The reusable fabrics are basically woven material of different fibres and their fibre blends such as 100% cotton, cotton/polyester blends and microfilament polyester. Wide range of fabrics that are single-use products called nonwoven, manufactured directly from fibres through mechanical and thermal means of fibre bonding techniques such as hydro entanglement, thermal bonding, stitch bonding and lamination [71].

Cotton and cotton/polyester blended fabric are traditional materials that are tightly woven focusing on wearer comfort but exhibiting large pore size that allows fluid and pathogens to transmit through, until special barrier finishes or coating applied on the fabric. These fabrics are reusable and it can be washed and sterilized by either steam autoclave or dry heat method. The life of reusable surgical fabrics are limited to maximum of 50 washing and drying cycles depending on the manufacturer label specification to avoid severe lint fibre shedding [72]. The liquid repellent of large pore woven fabrics is achieved by hydrophobic finishes and hot roller pressing. Microfilaments of polyester are densely woven to provide reliable barrier effect with hydrophobic finish. Fluorocarbon based liquid repellent, silver and copper-based coating for antimicrobial property along with blood repellent and hydrophobic self-cleaning effects are available commercially [73]. However, if the finish is not permanent, in the subsequent washing cycles and it can be reapplied after certain wash cycles. Multilayer disposable fabrics are produced with micro pores which are breathable but resist the penetration of liquids like blood and body serum. Microorganisms and infectious pathogens can persist over a range of few weeks to months depending on fibre type and the atmospheric temperature and relative humidity. Highly humid climate can increase the chance of bacteria and virus to stay alive for a longer period. It was found that cotton material can develop more bacteria than plastic suits in isolation ward and the health worker gowns and aprons are the highly contaminating cloths in hospital environment [69]. According to the association of surgical technologists, surgical scrub (cloth) that is made of cotton material has limitations in terms of three factors despite its rich softness in touch and breathability. Firstly, the fabric has large pores as it cannot be woven densely which in turn allows dispersal of skin scales and body liquids to pass through. Secondly, cotton fabrics are highly subjected to catching fire and thirdly, it fails to meet lint shedding resistivity [74].

The material specification for surgical gowns anticipated to use in healthcare products are elaborated in ASTM F 2407–06 devised in the year of 2013 and ASTM F 1671/F 1671 M-13 specifies the test methods and standards for determining resistance of protective fabric for penetration of blood borne pathogens. Wearer thermal comfort of surgical gowns standard test method and measurement are specified in ASTM F 1868–17 whereas the compliance of surgical cloth in terms of mechanical properties such as seam strength, breaking strength and elongation, resistance against tear and vapour transmission rate on standards are specified in ASTM D 1683/D 1683 M-17(2018), D 5034–09(2017), D 5587–15(2019) and D 6701–16 standards respectively. The fabrics that are coating with functional materials must follow the procedure and test standards as per ASTM D 751–19 [75].

AAMI (Association for the Advancement of Medical Instrumentation) described the test procedure and standard with criteria of acceptance for medical protective suits. Based on the grade of barrier performance the test and acceptance criteria and categorized for surgical protective suits and materials into four levels of performance. The surgical gown should accomplish the Level 4 performance according to ASTM F 1670 and F 1671 which should exhibit no penetration for blood and virus through the fabric. The acceptance value for all the levels and test criteria is 4% [76] (Table 1).

Protective aprons are suits that cover front body being worn over the surgical gown or over the cloth and it is used if the gown or suit worn by health care workers does not exhibit resistance to penetration of infectious pathogens [26]. Tri-layer

Table 1 AAMI Levels of performance of protective suits and materials [77–79]

Barrier levels	Performance	Standards	Criteria
Level 1	Minimal protection against Impact penetration of water	AATCC TM 42	<4.5 g for Water Impact
Level 2	Low protection against penetration of liquid at spray impact and at Hydrostatic Pressure	AATCC TM 42	≤ 1.0 g for Spray Impact
		AATCC TM 127	≥ 20 cm for Hydrostatic Pressure
Level 3	Moderate protection against penetration of liquid at spray impact and at Hydrostatic Pressure	AATCC TM 42	≤ 1.0 g for Spray Impact
		AATCC TM 127	≥ 50 cm for Hydrostatic Pressure
Level 4	High protection against Liquid (Surrogate Blood) and Viral (Bacteriophage Phi-X174) Penetration	ASTM F 1670 and F 1671	2 psi for Impermeable

composite fabrics are developed for the application of surgical gowns using polypropylene spun bond and melt blown layers and spun lace (hydro entangled) fabric of polyester/rayon fibre blend. The three layers provide both functions of barrier performance and wearer comfort by revealing reduced continuous pores and pore size along with breathability. Melt blown polypropylene is a nonwoven fabric, manufactured using a melt blown extrusion process. The main raw material used for this manufacturing process is a molten polymer, which forms a web like structure. It has the following properties which are essential for a face mask, (i) breathability, (ii) washable, (iii) stretchable, (iv) good absorbency and (v) excellent filtration. With additional antimicrobial and water repellent finish, the barrier performance Level of 4 according to AAMI Standards [80].

10   Alcohol-Based Hand sanitisers

Washing the hand frequently may reduce the spread of the virus. The effective hand wash kills or removes the contagious pathogens. The hand sanitiser formulation based on alcohol have become popular than the traditional handwashing based on soap as per the guidelines from the healthcare community [81]. Alcohol-based hand sanitisers may contain ethanol, isopropyl alcohol, n-propanol, or a mixture of these with excipients and humectants, as well as water. The most popular and efficient solutions include alcohols in the range of 60–95% by volume. Excipients help stabilize the substance and prolong the time it takes for alcohol to evaporate, thus enhancing its biocidal activity. Humectants help avoid skin dehydration and excipients help stabilize the product and prolong the time it takes for alcohol to evaporate, thereby raising its biocidal activity [82]. Alcohol-based disinfectants are good in destroying the transient organisms quickly below 30 s and the alcohol-based disinfectants for hand sanitiser also resident to microflora. The use of iodophors, hexachlorophene, chlorhexidine and alcoholic chlorhexidine repeatedly results in good residual activity than the other alcohol-based disinfectants.

11 Hand Drying Tissues

Hand drying tissues are essential parts of every good hand hygiene routines. Bacteria may be easily extracted from the hands by cleaning them with soap and water or only water and drying them with paper towels [83]. When it comes to drying the hand to remove moisture from the hand. Cloth roller towels are not advised because they can degrade into common-use towels at the end of the roll, potentially transferring pathogens to clean hands. Friction has been shown to be a crucial component in hand drying for eliminating contaminants in several tests. The mechanical abrasive activity of drying with paper towels eliminated bacteria from washed hands. After using the paper towels, The microbiological research, showed that many microbes were moved from the hands to the paper towels [84].

12 Biohazard Bags

Biohazard pouches are paper or plastic bags intended for containing, transporting and disposing of medical and hazardous materials safely. Biohazard pouches are hermetically sealed with zip-lock fastenings and can be used to store biohazardous specimens in a clean, airtight space. These guards against possible contamination of laboratory and research materials, as well as protecting handlers from possibly harmful substances [85]. Biohazard bags also known as Ziploc bags are used in carrying to the laboratory for testing. A bio-bag is a red coloured bag used to store solid biohazardous or infectious waste. It is also known as a biohazard waste storage bag or a red bag. These bags are mostly used to distinguish hazardous waste from normal trash that must be sterilized [86, 87].

The sealed bag acts as a secure disposal solution that avoids recontamination of the different containers' growth media. The biohazard bags are disposed away in the regular garbage after autoclave. When the bag has been autoclaved/steam-processed, an indication patch on the bag turns black. The dark patch serves as simple external evidence that the bag has been sterilized and should not be opened [88].

13 Hand Gloves

COVID-19 virus stay alive on all surfaces and the half-life of SARS-CoV-2 virus on different materials depend on the environment and the nature of the material. On stainless steel surfaces, the median value of half-life of this virus is 5.6 h and on plastic it is 6.8 h. In humid environment, the virus can stay approximately for 10–12 days [89]. Human hand is the main medium of transmission of viruses and pathogens through direct contact as it touches many contaminated surfaces and increases the chance of contracting with infection. Doctors and world health organization strongly recommends everyone to wash their hands at regular interval of time in the process of fight against COVID-19. Health care workers need to wash their hands frequently using alkaline soaps and alcohol-based sanitisers, which affect their skin and cause damages too [90, 91]. Excessive usage of water and chemical agents in hand resulted [92] in pathophysiologic changes (disruption of epidermal barrier layer) and dermatologic effects (extreme dryness, irritation and allergic, etc.).

It is recommended that in outpatient to isolation wards, all physicians, health care workers, cleaners and including visitors to wear gloves as mandatory protective suit, where cleaners are provided with heavy duty gloves for their protection. The ECDC (European Centre for Disease Prevention and Control) and CDC (Centre for Disease Control and Prevention) stated that prolonged use of medical gloves can increase the risk of self-contamination as it involves involuntary touch on face after working with desks, telephones, mobiles and computers. In connection to this ECDC and CDC have recently released the guidelines for medical professionals, health care staff and general community related to the use of gloves, as given below [93];

1. Not to use sterile gloves beyond the recommended shelf-life, mentioned by the manufacturer.
2. Non-sterile disposable gloves should be used on priority where it poses the chance of contacting epidemic patients and their blood samples.
3. Use medical grade gloves in places where a person is not exposed to pathogens.
4. Extending the period of usage of disposable gloves with proper hand sanitizing after removal of gloves.

Some precautionary measures to be taken care while using hand gloves, such as avoiding few hand lotions and abrasive chemicals that can damage latex type gloves. And one should avoid touching mucosal membranes such as the eye, nose and mouth with contaminated worn gloves as it increases the risk of contracting infection at a great level [94]. The standardized specifications of different types of medical grade examination gloves for manufacturing nitrile rubber, latex rubber, polyvinylchloride and polychloroprene gloves material are prescribed in ASTM D6319-19, D3578-19, D5250-19 and D6977-19 modules respectively whereas the standardized specifications for surgeon gloves are prescribed in ASTM D3577-19 [75].

14 Eye Protection and Face Shields

The surveys from US and China healthcare facilities have shown that, 70% of health care workers who worked closely with patients were affected and caused to death as they unaware of eye or face protection despite their complete protection with gown and respirator face masks [95]. World health organization made eye protection or facial protection as imperative parts of PPE suit in addition to respiratory masks and gloves to prevent contact, droplet transmission and airborne transmission of pathogens [96]. Goggles are used as equipment worn around eyes, whereas face shields and masks cover from forehead to chin, are used for protecting parts such as eyes, nose and mouth and preventing airborne and droplet transmission [97].

Eye protection shields include three major parts such as frame, visor and suspension system. Frame is the holding part and fixes the shield onto the head portion of the person or wearer and it can be made of plastics with adjusting metal clips. Visor is the main protecting part with 100% light transmitting capacity can also be called window or lens. It comes with anti-fog, antiglare, antistatic, scratch resistance and UV protection properties. The suspension system is to reach the adjustability to the wearer fit. This is made of plastics, works with ratchet and pin-lock mechanisms or Velcro attachments or elastic straps [98, 99]. Eye protective goggles or face shields along with mask are preferred places where splashes, spray and splatter of blood droplets are potential sources of infection transmission. ANSI/ISEA Z87.1—2015 Standard state the product specifications such as material design, performance and quality for manufacturing goggles, face shields, spectacles and helmets for health care officials [100].

15 Conclusion

On December 31 of 2019, the Chinese authorities sent an alert message about the outbreak of a strange strain of coronavirus (subsequently named SARS-CoV-2) causing multiple illnesses. Subsequently, on 30 January 2020, WHO declared the outbreak a Public Health Emergency of International Concern and then as a pandemic on 11 March 2020. SARS virus, a type of coronavirus that has a size of 100 nm can easily pass through many barriers, like influenza virus (80–120 nm). Though the size of the new COVID-19 virus is currently unknown, human coronaviruses are generally about 125 nm well below the cut-off values available for mechanical respirators. However, viruses often travel on top of larger carrier molecules—like globs of mucus—making it easier to filter them, using PPEs including masks. Judicious selection of filter materials coupled with certain degree of comfort to the wearer would be, essentially, the choice of many who suffer due to the diseases caused by COVID and who are involved in providing treatments to the affected ones. COVID-19 pandemic situation makes us wear face mask compulsory and thus increase the demand for such materials. Most of the face masks are based on synthetic materials with plastic contents. This is, obviously, expected to introduce the challenge in the disposal of such face masks, which could be the area for research many.

References

1. Chaib F (2020) Shortage of personal protective equipment endangering health workers worldwide. In: World Health Organization. https://www.who.int/news/item/03-03-2020-shortage-of-personal-protective-equipment-endangering-health-workers-worldwide. Accessed on 26 Mar 2021
2. Murray OM, Bisset JM, Gilligan PJ et al (2020) Respirators and surgical facemasks for COVID-19: implications for MRI. Clin Radiol 75:405–407. https://doi.org/10.1016/j.crad.2020.03.029
3. Wilder-Smith A, Freedman DO (2020) Isolation, quarantine, social distancing and community containment: pivotal role for old-style public health measures in the novel coronavirus (2019-nCoV) outbreak. J Travel Med 27:1–4. https://doi.org/10.1093/jtm/taaa020
4. Lin Y-H, Liu C-H, Chiu Y-C (2020) Google searches for the keywords of "wash hands" predict the speed of national spread of COVID-19 outbreak among 21 countries. Brain Behav Immun 87:30–32. https://doi.org/10.1016/j.bbi.2020.04.020
5. Chintalapudi N, Battineni G, Amenta F (2020) COVID-19 virus outbreak forecasting of registered and recovered cases after sixty day lockdown in Italy: a data driven model approach. J Microbiol Immunol Infect 53:396–403. https://doi.org/10.1016/j.jmii.2020.04.004
6. Elachola H, Ebrahim SH, Gozzer E (2020) COVID-19: facemask use prevalence in international airports in Asia, Europe and the Americas, March 2020. Travel Med Infect Dis 35:101637. https://doi.org/10.1016/j.tmaid.2020.101637
7. Yang P, Seale H, Raina MacIntyre C et al (2011) Mask-wearing and respiratory infection in healthcare workers in Beijing, China. Brazilian J Infect Dis 15:102–108. https://doi.org/10.1016/S1413-8670(11)70153-2
8. Leung CC, Lam TH, Cheng KK (2020) Mass masking in the COVID-19 epidemic: people need guidance. Lancet 395:945. https://doi.org/10.1016/S0140-6736(20)30520-1
9. Elachola H, Assiri AM, Memish ZA (2014) Mass gathering-related mask use during 2009 pandemic influenza A (H1N1) and Middle East respiratory syndrome coronavirus. Int J Infect Dis 20:77–78. https://doi.org/10.1016/j.ijid.2013.12.001
10. Barasheed O, Alfelali M, Mushta S et al (2016) Uptake and effectiveness of facemask against respiratory infections at mass gatherings: a systematic review. Int J Infect Dis 47:105–111. https://doi.org/10.1016/j.ijid.2016.03.023
11. Wu H, Huang J, Zhang CJP et al (2020) Facemask shortage and the novel coronavirus disease (COVID-19) outbreak: reflections on public health measures. EClinicalMedicine 21:100329. https://doi.org/10.1016/j.eclinm.2020.100329
12. Bozhilova M (2018) Difference between parasite and virus. Available via DIALOG. http://www.differencebetween.net/science/health/difference-between-parasite-and-virus/. Accessed 28 Mar 2021
13. They differ (2018) Difference between DNA and RNA. https://theydiffer.com/difference-between-dna-and-rna/. Accessed 27 Mar 2021
14. Lovelace B (2020) WHO officials scramble to measure size of coronavirus epidemic: How big is the iceberg? https://www.cnbc.com/2020/02/13/who-officials-scramble-to-measure-size-of-coronavirus-epidemic-how-big-is-the-iceberg.html. Accessed 28 Mar 2021

15. Beckman (2020) What are the differences between RNA and DNA viruses used in immunotherapy? https://mybeckman.in/support/faq/research/rna-and-dna-viruses-in-immunotherapy. Accessed 28 Mar 2021

16. Harold G (2019) DNA viruses vs RNA viruses. In: Diffzi. https://diffzi.com/dna-viruses-vs-rna-viruses. Accessed 28 Mar 2021

17. Lakna (2018) Difference between positive and negative sense RNA virus. https://pediaa.com/difference-between-positive-and-negative-sense-rna-virus. Accessed 28 Mar 2021

18. Edwards E, Miller SG (2020) The world learned about a new coronavirus 5 months ago. Here's what we now know about COVID-19. In: CNBC news. https://www.who.int/publications/i/item/WHO-2019-nCoV-clinical-2021-1. Accessed 28 Mar 2021

19. Rossen J (2020) Does wearing a face mask really help protect against coronavirus and other illnesses? https://www.mentalfloss.com/. Accessed 28 Mar 2021

20. Larsen S (2021) Coronavirus is an RNA Virus|COVID-19 and the essential nutrition to protect yourself. https://drsarahlarsen.com/. Accessed 28 Mar 2021

21. Uslocationsca (2019) Can commercial air filtration systems remove viruses in the atmosphere? In: Camfil. https://cleanair.camfil.us/2019/11/22/can-commercial-air-filtration-systems-remove-viruses-in-the-atmosphere/. Accessed 28 Mar 2021

22. HNGN (2020) Coronavirus: can it be transmitted via air-conditioning outlets and ducts? In: HNGN. https://www.hngn.com/articles/228319/20200309/coronavirus-transmitted-via-air-conditioning-outlets-ducts.htm. Accessed 28 Mar 2021

23. U.S Food & Drug Administration (2020) N95 respirators, surgical masks, and face masks. In: U.S Food Drug Adm. https://www.fda.gov/medical-devices/personal-protective-equipment-infection-control/n95-respirators-surgical-masks-and-face-masks. Accessed 27 Mar 2021

24. Honeywell (2020) N95 masks explained. In: Honeywell. https://www.honeywell.com/us/en/news/2020/03/n95-masks-explained. Accessed 27 Mar 2021

25. Bessesen MT, Savor-Price C, Simberkoff M et al (2013) N95 respirators or surgical masks to protect healthcare workers against respiratory infections: are we there yet? Am J Respir Crit Care Med 187:904–905. https://doi.org/10.1164/rccm.201303-0581ED

26. WHO (2020) Rational use of personal protective equipment for coronavirus disease 2019 (COVID-19)—interim guidance. WHO interim Guid 2019:1–7

27. Wells WF (1934) On air-borne infection*. Am J Epidemiol 20:611–618. https://doi.org/10.1093/oxfordjournals.aje.a118097

28. Eames I, Shoaib D, Klettner CA, Taban V (2009) Movement of airborne contaminants in a hospital isolation room. J R Soc Interface 6:S757–S766. https://doi.org/10.1098/rsif.2009.0319.focus

29. Fiegel J, Clarke R, Edwards DA (2006) Airborne infectious disease and the suppression of pulmonary bioaerosols. Drug Discov Today 11:51–57. https://doi.org/10.1016/S1359-6446(05)03687-1

30. Nicas M, Nazaroff WW, Hubbard A (2005) Toward understanding the risk of secondary airborne infection: emission of respirable pathogens. J Occup Environ Hyg 2:143–154. https://doi.org/10.1080/15459620590918466

31. Chao CYH, Wan MP, Morawska L et al (2009) Characterization of expiration air jets and droplet size distributions immediately at the mouth opening. J Aerosol Sci 40:122–133. https://doi.org/10.1016/j.jaerosci.2008.10.003

32. Gobalakrishnan M, Saravanan D (2020) Antimicrobial activity of gloriosa superba, cyperus rotundus and pithecellobium dulce with different solvents. Fibres Text East Eur 28:67–71. https://doi.org/10.5604/01.3001.0014.0937

33. Czajka R (2005) Development of medical textile market. Fibres Text East Eur 13:13–15

34. Gobalakrishnan M, Saravanan D (2017) Antimicrobial activity of coleus ambonicus herbal finish on cotton fabric. Fibres Text East Eur 25:106–109. https://doi.org/10.5604/01.3001.0010.2854

35. Sumithra M, Vasugi Raaja N (2012) Micro-encapsulation and nano-encapsulation of denim fabrics with herbal extracts. Indian J Fibre Text Res 37:321–325

36. Huang X, Netravali A (2009) Biodegradable green composites made using bamboo micro/nano-fibrils and chemically modified soy protein resin. Compos Sci Technol 69:1009–1015. https://doi.org/10.1016/j.compscitech.2009.01.014
37. International cotton advisory committee (2013) World apparel fiber consumption survey
38. Gobalakrishnan M, Saravanan D, Das S (2020) Sustainable finishing process using natural ingredients. In: Muthu SS (ed) Sustainability in the textile and apparel industries. Springer Nature, Switzerland
39. Meléndez PA, Capriles VA (2006) Antibacterial properties of tropical plants from Puerto Rico. Phytomedicine 13:272–276. https://doi.org/10.1016/j.phymed.2004.11.009
40. Chandrasekaran K, Ramachandran T, Vigneswaran C (2012) Effect of medicinal herb extracts treated garments on selected diseases. Indian J Tradit Knowl 11:493–498
41. Tao C, Wu J, Liu Y et al (2018) Antimicrobial activities of bamboo (Phyllostachys heterocycla cv. Pubescens) leaf essential oil and its major components. Eur Food Res Technol 244:881–891. https://doi.org/10.1007/s00217-017-3006-z
42. Kweon MH, Hwang HJ, Sung HC (2001) Identification and antioxidant activity of novel chlorogenic acid derivatives from bamboo (Phyllostachys edulis). J Agric Food Chem 49:4646–4655. https://doi.org/10.1021/jf010514x
43. Afrin T, Tsuzuki T, Kanwar RK, Wang X (2012) The origin of the antibacterial property of bamboo. J Text Inst 103:844–849. https://doi.org/10.1080/00405000.2011.614742
44. Hussain U, Bin YF, Usman F et al (2015) Comfort and mechanical properties of polyester/bamboo and polyester/cotton blended knitted fabric. J Eng Fiber Fabr 10:61–69. https://doi.org/10.1177/155892501501000207
45. Sarkar AK, Appidi S (2009) Single bath process for imparting antimicrobial activity and ultraviolet protective property to bamboo viscose fabric. Cellulose 16:923–928. https://doi.org/10.1007/s10570-009-9299-8
46. Xiao Z, Zhang Q, Dai J et al (2020) Structural characterization, antioxidant and antimicrobial activity of water-soluble polysaccharides from bamboo (Phyllostachys pubescens Mazel) leaves. Int J Biol Macromol 142:432–442. https://doi.org/10.1016/j.ijbiomac.2019.09.115
47. Rajendran R, Radhai R, Kotresh TM, Csiszar E (2013) Development of antimicrobial cotton fabrics using herb loaded nanoparticles. Carbohydr Polym 91:613–617. https://doi.org/10.1016/j.carbpol.2012.08.064
48. Joshi M, Ali SW, Purwar R, Rajendran S (2009) Ecofriendly antimicrobial finishing of textiles using bioactive agents based on natural products. Indian J Fibre Text Res 34:295–304
49. Tcharkhtchi A, Abbasnezhad N, Zarbini Seydani M et al (2021) An overview of filtration efficiency through the masks: mechanisms of the aerosols penetration. Bioact Mater 6:106–122. https://doi.org/10.1016/j.bioactmat.2020.08.002
50. Sickbert-Bennett EE, Samet JM, Clapp PW et al (2020) Filtration efficiency of hospital face mask alternatives available for use during the COVID-19 pandemic. JAMA Intern Med 27516:1–6. https://doi.org/10.1001/jamainternmed.2020.4221
51. COVID-19 clinical management: living guidance (2021) World Health Organization. https://www.who.int/publications/i/item/WHO-2019-nCoV-clinical-2021-1. Accessed 27 Mar 2021
52. Dondorp AM, Hayat M, Aryal D et al (2020) Respiratory support in COVID-19 patients, with a focus on resource-limited settings. Am J Trop Med Hyg 102:1191–1197. https://doi.org/10.4269/ajtmh.20-0283
53. Ranney ML, Griffeth V, Jha AK (2020) Critical supply shortages—the need for ventilators and personal protective equipment during the Covid-19 pandemic. N Engl J Med 382:e41. https://doi.org/10.1056/NEJMp2006141
54. Isinnova (2020) Easy—Covid19 ENG. https://isinnova.it/easy-covid19-eng. Accessed 27 Mar 2021
55. Ventilators and ventilator accessories EUAs (2020) FDA, https://www.fda.gov/medical-devices/coronavirus-disease-2019-covid-19-emergency-use-authorizations-medical-devices/ventilators-and-ventilator-accessories-euas. Accessed 27 Mar 2021

56. US Food & Drug Administration (2020) EUA Letter of Authorization - Ventilators, anesthesia gas machines modified for use as ventilators, and positive pressure breathing devices modified for use as ventilators (collectively referred to as "ventilators"), ventilator tubing connectors, and vent. 2019:1–15

57. Cao G, Awbi H, Yao R et al (2014) A review of the performance of different ventilation and airflow distribution systems in buildings. Build Environ 73:171–186. https://doi.org/10.1016/j.buildenv.2013.12.009

58. Advincula RC, Dizon JRC, Chen Q et al (2020) Additive manufacturing for COVID-19: devices, materials, prospects, and challenges. MRS Commun 10:413–427. https://doi.org/10.1557/mrc.2020.57

59. King WP, Amos J, Azer M et al (2020) Emergency ventilator for COVID-19. PLoS ONE 15: e0244963. https://doi.org/10.1371/journal.pone.0244963

60. Slutsky AS, Ranieri VM (2013) Ventilator-induced lung injury. N Engl J Med 369:2126–2136. https://doi.org/10.1056/NEJMra1208707

61. Vieillard-Baron A, Rabiller A, Chergui K et al (2005) Prone position improves mechanics and alveolar ventilation in acute respiratory distress syndrome. Intensive Care Med 31:220–226. https://doi.org/10.1007/s00134-004-2478-z

62. Nacharaju D, Menzel W, Fontaine E et al (2020) Three-dimensional printed ventilators: a rapid solution to coronavirus disease 2019–induced supply-chain shortages. Crit Care Explor 2:e0226. https://doi.org/10.1097/CCE.0000000000000226

63. Wang J, Du G (2020) COVID-19 may transmit through aerosol. Irish J Med Sci 189:1143–1144. https://doi.org/10.1007/s11845-020-02218-2

64. Rodrigues-Pinto R, Sousa R, Oliveira A (2020) Preparing to perform trauma and orthopaedic surgery on patients with COVID-19. J Bone Jt Surg 102:946–950. https://doi.org/10.2106/JBJS.20.00454

65. Liang ZC, Chong MSY, Sim MA et al (2020) Surgical considerations in patients with COVID-19. J Bone Jt Surg 102:e50. https://doi.org/10.2106/JBJS.20.00513

66. Behera BK, Arora H (2009) Surgical gown: a critical review. J Ind Text 38:205–231. https://doi.org/10.1177/1528083708091251

67. Garner JS, Simmons BP (1983) Guideline for isolation precautions in hospitals. Infect Control 4:245–325. https://doi.org/10.1017/s0195941700003763

68. Stewart CL, Thornblade LW, Diamond DJ et al (2020) Personal protective equipment and COVID-19. Ann Surg 272:e132–e138. https://doi.org/10.1097/SLA.0000000000003991

69. Kilinc FS (2015) A review of isolation gowns in healthcare: fabric and gown properties. J Eng Fiber Fabr 10:155892501501000. https://doi.org/10.1177/155892501501000313

70. Mitura K, Myśliwiec P, Rogula W et al (2020) Guidelines for the management of surgical departments in non-uniform hospitals during the COVID-19 pandemic. Polish J Surg 92:48–59. https://doi.org/10.5604/01.3001.0014.1039

71. Abreu MJ, Silva ME, Schacher L, Adolphe D (2003) Designing surgical clothing and drapes according to the new technical standards. Int J Cloth Sci Technol 15:69–74. https://doi.org/10.1108/09556220310461178

72. Whyte W, Bailey P, Hamblen D et al (1983) A bacteriologically occlusive clothing system for use in the operating room. J Bone Joint Surg Br 65B:5025–5026. https://doi.org/10.1302/0301-620X.65B4.6874723

73. Suryaprabha T, Sethuraman MG (2017) Fabrication of copper-based superhydrophobic self-cleaning antibacterial coating over cotton fabric. Cellulose 24:395–407. https://doi.org/10.1007/s10570-016-1110-z

74. Association of surgical technologists (2008) AST standards of practice for surgical attire, surgical scrub, hand hygiene and hand washing. https://www.ast.org/uploadedfiles/main_site/content/about_us/standard_surgical_attire_surgical_scrub.pdf. Accessed 26 Mar 2021.

75. ASTM (2020) ASTM standards & COVID-19. https://www.astm.org/COVID-19. Accessed 26 Mar 2021

76. AAMI (2020) Personal protective equipment. In: AAMI. https://munglobal.com.au/resources/knowledge-base/personal-protective-equipment/aami-level-standards-for-gowns/. Accessed 25 Mar 2021

77. Diana A Wyman (2020) Impact penetration testing for textiles part 1. In: AATCC. https://www.aatcc.org/impact-penetration-testing-for-textiles/. Accessed 26 Mar 2021

78. AATCC 127: water resistance: hydrostatic pressure test from American Association of Textile chemists and colorists

79. ASTM F1670/F1670M: standard test method for resistance of materials used in protective clothing to penetration by synthetic blood from ASTM

80. Moradi F, Ahmadi MS, Mashroteh H (2019) Development of tri-layer breathable fluid barrier nonwoven fabrics for surgical gown applications. J Text Inst 110:1545–1551. https://doi.org/10.1080/00405000.2019.1606378

81. Pittet D, Allegranzi B, Boyce J (2009) The world health organization guidelines on hand hygiene in health care and their consensus recommendations. Infect Control Hosp Epidemiol 30:611–622. https://doi.org/10.1086/600379

82. Donnell Gerald, Denver Russell A (1999) Antiseptics and disinfectants: activity, action, and resistance. Clin Microbiol Rev 12:147–179

83. Boyce JM, Pittet D (2002) Guideline for hand hygiene in health-care settings. Am J Infect Control 30:S1–S46. https://doi.org/10.1067/mic.2002.130391

84. Sprunt K, Redman W, Leidy G (1973) Antibacterial effectiveness of routine hand washing. Pediatrics 52:264–271

85. Levins C (2018) Who uses biohazard pouches? https://www.airseacontainers.com/blog/who-uses-biohazard-pouches/. Accessed 26 Mar 2021

86. Amukele TK, Hernandez J, Snozek CL et al (2017) Drone transport of chemistry and hematology samples over long distances. Am J Clin Pathol 148:427–435. https://doi.org/10.1093/ajcp/aqx090

87. Brenner B (2016) Bio bags—when to use them. https://www.medprodisposal.com/biohazard-waste-disposal/when-to-use-bio-bag/. Accessed 26 Mar 2021

88. Fax BIO (2016) Biohazard Disposal Bags. Flinn Sci 10127:1–2

89. van Doremalen N, Bushmaker T, Morris DH et al (2020) Aerosol and surface stability of SARS-CoV-2 as compared with SARS-CoV-1. N Engl J Med 382:1564–1567. https://doi.org/10.1056/NEJMc2004973

90. Alanagreh L, Alzoughool F, Atoum M (2020) The human coronavirus disease COVID-19: its origin, characteristics, and insights into potential drugs and its mechanisms. Pathogens 9:331. https://doi.org/10.3390/pathogens9050331

91. Cavanagh G, Wambier CG (2020) Rational hand hygiene during the coronavirus 2019 (COVID-19) pandemic. J Am Acad Dermatol 82:e211. https://doi.org/10.1016/j.jaad.2020.03.090

92. Beiu C, Mihai M, Popa L et al (2020) Frequent hand washing for COVID-19 prevention can cause hand dermatitis: management tips. Cureus. https://doi.org/10.7759/cureus.7506

93. Centre for Disease Control and Prevention (2020) Strategies for optimizing the supply of disposable medical gloves. https://www.cdc.gov/coronavirus/2019-ncov/hcp/ppe-strategy/gloves.html. Accessed 27 Mar 2021

94. Smereka J, Szarpak L (2020) The use of personal protective equipment in the COVID-19 pandemic era. Am J Emerg Med 38:1529–1530. https://doi.org/10.1016/j.ajem.2020.04.028

95. Chan JF-W, Yuan S, Kok K-H et al (2020) A familial cluster of pneumonia associated with the 2019 novel coronavirus indicating person-to-person transmission: a study of a family cluster. Lancet 395:514–523. https://doi.org/10.1016/S0140-6736(20)30154-9

96. Marra AR, Edmond MB, Popescu SV, Perencevich EN (2020) Examining the need for eye protection for coronavirus disease 2019 (COVID-19) prevention in the community. Infect Control Hosp Epidemiol 42:646–647. https://doi.org/10.1017/ice.2020.314

97. World Health Organization (2020) Infection prevention and control during health care when novel coronavirus (nCoV) infection is suspected. Interim Guid. https://www.who.int/publications/i/item/10665-331495. Accessed January 2020.

98. Roberge RJ (2016) Face shields for infection control: a review. J Occup Environ Hyg 13:235–242. https://doi.org/10.1080/15459624.2015.1095302

99. Chu DK, Akl EA, Duda S et al (2020) Physical distancing, face masks, and eye protection to prevent person-to-person transmission of SARS-CoV-2 and COVID-19: a systematic review and meta-analysis. Lancet 395:1973–1987. https://doi.org/10.1016/S0140-6736(20)31142-9

100. Baugh L (2015) The PPE industry responds to COVID-19. In: International Safety Equipment Association. https://safetyequipment.org/standard/ansiisea-z87-1-2015/. Accessed 27 Mar 2021

Design and Development of a Three Layered Surgical Mask for Healthcare Professionals Against COVID-19

V. Parthasarathi and M. Parthiban

Abstract The World Health Organization's (WHO) guidelines regarding prevention and control of the COVID-19 outbreak recommends respiratory hygiene and the use of personal protective equipment. Three layered antiviral mask has been developed by combining polyester nonwoven as an outer, polytetrafluoroethylene (PTFE) as a middle and viscose nonwoven as an inner layer. All three layers of fabrics were fused as laminate to develop the mask to protect healthcare professionals from coronavirus. All the surgical mask tests were performed to make compliance with global standards. Primarily viral penetration and bacterial filtration efficiency was performed according to ASTM standard F1671 and F2100, respectively. It has been observed that the mean flow pore diameter of single layer polyester surgical mask, dual layer of polyester and PTFE film surgical mask and three layered surgical mask was 0.33, 0.30 and 0.39 µm. The developed three layered fabric with polyester as an outer layer, microporous PTFE film as a middle layer and viscose nonwoven fabric as an inner layer has passed the viral penetration test, which confirms that the mask has protection against coronavirus. Paired t'-test analysis of the tensile strength in machine and cross directions shows that there is significant difference at 95% confidence interval. Single layer polyester nonwoven surgical mask failed viral penetration test as the average pore diameter is 0.33 µm which is more than size of ΦX174 bacteriophage virus as 0.2 µm and in the absence of PTFE film. Dual and three layered surgical mask passed viral penetration analysis as it has smallest pore diameter is 0.16 and 0.19 µm, respectively. PTFE film allows water vapor to pass through. In the current pathetic and demanding situation, the three layered surgical mask was having barrier against coronavirus which facilitate healthcare care professionals from cross-contamination and protecting them from current pandemic.

Keywords Antiviral · COVID-19 · Nonwoven · PTFE · PET and surgical mask

V. Parthasarathi · M. Parthiban (✉)
Department of Fashion Technology, PSG College of Technology, Peelamedu, Coimbatore 641 004, India

S. S. Muthu (ed.), *COVID-19*, Environmental Footprints and Eco-design of Products and Processes, https://doi.org/10.1007/978-981-16-3856-5_8

221

1   Introduction

COVID-19 is an extraordinary worldwide danger which causes respiratory issues which may result deadly. Since COVID-19 is an illness spread by vaporized drops from the aviation route of tainted patients, it is announced as a pandemic by the World Health Organization on March 11, 2020. For handling this COVID-19 pandemic, endeavours are taken to forestall transmission of infection to bleeding edge medical service, labours who are taking counter measures to treat the influenced people. Apart from the services, we need a stringent measure for protecting ourselves through individual personal protective equipment. Currently, PPE has become an international concern [1]. In the present pandemic situation, one should not be able to spread the contamination that may influence the resulting time of re-organizing the ordinary existence of public [2]. Hence, PPE is currently the need of the hour [3].

As indicated by Wuetal ponders, coronavirus is principally imparted by methods for respiratory drops and could be transmitted between individuals. Symptoms of fever, hack, exhaustion, cerebral agony and haemoptysis are the indicators of the serious illness caused by coronavirus. In the growing need of different far and wide facemask wearing game plan strategies, insufficiency of facemask production would occur. According to World Health Organization (WHO), three layer protective materials are crucial as far as clinical procedures and providers are concerned [4]. The present pandemic conditions may pave a way for the inadequacy of filtering face cover for clinical consideration professionals. From research studies, it was found that that the surface execution is changed when surfaces at first interface with a fluid which may collaborate to decrease the resistance of material to liquid invasion [5]. Personal protective equipments should give protection limit against the transmission of microorganisms and fluids to restrict threat of strike-through and for the patient's security [6].

In the public's view of medical service experts, outfits assume a significant part. The reliability is decided by the garments worn by the patients. Socially developed ideas of neatness that bring about unattainable assumptions might be supported by the tone and plan of regalia. These should be cultivated in a financially savvy way. Sickness to their hosts is brought about by irresistible specialists which are natural specialists [7]. In medical service sector, more irresistible specialists are available. The patients and medical service personnel are probably wellsprings of irresistible specialists and the most well-known vulnerable hosts [8]. The medical association recommended that the textures utilized for veil and curtains should oppose fluid transmission, scraped spot and penetrates and limit entry of microbes from non-clean to clean areas. Neely and Maley (2000) suggested that the accompanying five emergency clinic materials, for example, 100% cotton, 100% cotton terry towel, polyester cotton mix scour suit, 100% polyester wrap and 100% polypropylene sprinkle aprons could be recommended for medical applications [9]. Both texture and non-texture materials could be utilized to make careful cleans [10]. The investigation was mainly intended to check the logical proof of wearing cleans depends on the materials they are made of as per the medical procedures [11].

Because of the predominance of Covid in the patient populace, the obstruction adequacy of defensive careful veils and cover has acquired significance [12]. Utilizing standard research facility conditions, which are not quite the same as the conditions experienced in the careful region, a large portion of the careful cover textures are tried and categorized [13]. Wei Huang & Karen studies inferred that antimicrobial with fluoro-substance repellent completions are applied to the two nonwoven textures, for example, polypropylene spun-fortified/dissolve blown/ spun-reinforced texture and a wood mash/polyester spun-laced fabric [14].

The significance of shielding medical care labours from body liquids has additionally been perceived. The bleeding edge medical services labours that are liable for controlling the careful climate ought to have the specialized, functional and predictable information on assessing the obstruction properties of careful clothing. There is no uniform code or rule that characterizes the favoured strategy or strategies for assessment, while numerous research facilities have, by need, settled upon a set number of test techniques to assess the fluid infiltration obstruction or repellency of a hindrance material [15].

To assess the overall performance, there are four important tests that could be performed to substantiate the same, namely spray impact penetration test; hydro-static head test; ASTM F1670 blood obstruction and ASTM F1671 viral hindrance testing. To decide, if the item is defensive or non-defensive, normally the spray impact test is performed [16]. The degree of security from 1 to 3 is demonstrated by the hydrostatic pressure test results. ASTM F1670 and ASTM F1671 tests need to be performed for those items which are needed to be completely impervious. Longer hospitalization of patients with supporting medical services, additional therapies, and anti-infection agents requires additional expenses. The emergency clinic stay of patients with a medical services-related disease is assessed to be 2.5 occasions longer than that of patients without such a contamination, and the expense of therapy is commonly higher [17].

Both reusables and disposables are better in various ecological effects. The variables to improve both reusables and dispensable frameworks are hard to eval-uate as enormous scope investigations of solace, assurance, or economics [18]. The main aim is to create and reuse these items are reflected in the contemporary examinations of reusable and single utilize preoperative materials like careful veils and curtains. Reusable and expendable covers and curtains are made by utilizing manufactured lightweight textures with a serious estimating to fulfil new guidelines for clinical labours and patient insurance. A per the life cycle ecological exami-nations which depends on numerous science, reusable careful veils and window hangings showed significant maintainability merits past a similar expendable item in normal asset energy scopes of 200–300%, water 250–330%, carbon impression 200–300%, unpredictable organics, and strong squanders 750%.

The ecological advantage of reusable careful veils and window hangings to medical care manageability programmes is significant for this industry as any remaining components like expense, security and solace level are sensibly comparative. The assertion demonstrates that reusables are better in some natural products, and disposables are better in other ecological effects is not, at this point substantial. The variables to improve both reusables and expendable frameworks are hard to evaluate as enormous scope investigations of solace, assurance, or financial aspects have not been effectively sought after in the last 5–10 years. As per OSHA rules for fitting utilization of individual defensive hardware (PPE), for example, face shields or covers, gloves, research centre covers and eye assurance and mouthpieces, pocket veils or other ventilation gadgets, there should be a global openness, liberated from cost to the employee [19].

The most probable wellsprings of irresistible specialists and furthermore the most widely recognized powerless hosts are patients and medical care personnel. Guests and people under medical care may be in the danger of both contamination and transmission [9]. The well-being of Chinese has to be treated with defined specialists to control the illness arrived at a resolution that the presumed sickness influenced individuals ought to be kept in seclusion and to screen them and their nearby contacts, gather clinical tried information from the tainted patients and advancement of therapy procedures [20]. For covers to be tested by body liquids like blood, sweat and different fluids like liquor or iodine, the chance is available in the idea of exercises in the careful climate. Just refined water as the test fluid is utilized in some standard testing techniques for setting up the defensive execution of careful veil textures. According to the insights of World Health Organization announced, 34 million individuals endured in human immune-deficiency infection worldwide [21].

Careful site infection (SSI) rates stay high inspite of innovative progressions in the careful focus region and more prominent information on clinic disease hazard factors. Apart from the fact that numerous SSI avoidance measures are suggested, the rules for the anticipation of careful site contamination of the centres for disease control and counteraction (CDC) of the USA, distributed in 1999, had been generally recognized. Inborn elements which are identified with patient, age, sort of a medical procedure, base pathology, related pathologies among different viewpoints and outward factors alluding to help strategies, medical procedure strategy, pre-medical procedure arrangement, climate, careful clothing and anti-infection prophylaxis had been considered [22].

Careful cover ought to be made out of fluid evidence texture to shield the blood-borne irresistible organisms from entering through the texture. Careful veil should give solace to the wearer by permitting warmth and mugginess to trade among inside and outside of the body. At the end of the day, the texture ought to be obstruction safe is most susceptible to wear. Hence, the texture should oppose the entrance of fluids, especially blood and simultaneously be clean, breathable, adaptable and inexpensive [23]. HIV and hepatitis B and hepatitis C infections in the patient populace have incited the use of careful covers as a hindrance. Attire assumes a fundamental and substantial part in giving physiological solace and furthermore in securing the human body. Defensive attire for natural perils is utilized in medical services area. Law implementation and morgue labourers need this defensive apparel as they need to shield themselves from different dangers in the environment [24].

Dangers of patients are defilement from both endogenous and exogenous microorganisms, and danger of medical services labourers are tainting from different blood-borne microbes because of word-related openness to patient blood and body fluids [25]. For the government assistance of medical care labourers just as their patients, the norms for HCWU are vital. Wong et al. found that monitoring the risks and wearing proper defensive boundaries can change the level of insurance for medical services staff essentially. Medical care offices have alternatives to choose items offering various degrees of security dependent on the perils during the surgeries, since surgeries include fluctuating degrees of sprinkling, inclining and pressing factor. Suggestions or commands have been made by a few associations on the most proficient method to shield HCW just as patients from openness to blood-borne microbes and bacteria [26]. In 1971, OSHA was set up as a division of the branch of work to secure labourers' wellbeing, forestall injury and save lives. OSHA suggests that when a worker has a potential for openness of the work, fitting defensive garments should be worn to frame a successful obstruction. Contingent on the word-related assignment and the level of likely openness, the kind of attire required is to be chosen. Defensive dress should be worn to forestall the worker basic apparel from pollution if the garments are possibly filthy from blood or other conceivably irresistible materials. At the point when labourers could get contaminated through sprinkling or splashing of blood or other conceivably irresistible materials, liquid safe dress should be worn. A particular defensive sort of boundary attire is required as a bigger volume of blood, and other possibly irresistible materials are related with crafted by the HCW. OSHA likewise suggests that towards the finish of the work move, the tainted uniform ought to be eliminated. A defiled uniform ought to be left at the work zone for cleaning, washing, or potentially arranging and ought not to be taken home [27].

An assortment of materials including customary polyester/cotton just as more present day textures, both woven and nonwoven, is utilized to make theatre veils and window hangings. Clients should pick among reusable and single use choices. European Committee for Standardization (CEN) distributed European Standards which are expected to help the fundamental prerequisites of European Union (EU) mandates. Theatre curtains and covers are the initial segment of another European Standard [28]. The prerequisites for producer cycle and items are managed in the primary part of European norm. The different testing strategies for properties, for example, protection from wet microbial infiltration have become later areas as per European norm. Quality norms in the key execution zones as depicted above as protections from bacterial and fluid entrance just as elastic and blasting strength are itemized in part one of European Standard [29].

It is a reasonable ramification that some sort of testing of qualification for reason between employments of reusable veils and curtains is implied by adjusting to this norm. In those pieces of standard that followed section one, the specific idea of such testing was definite [30]. Instances of testing incorporate visual investigation for net flaws like openings, just as pressing factor tests for wet and dry bacterial entrance and blasting strength have been examined. Medical services-related diseases (HAI), bleakness, mortality and delayed hospitalization are essential weight of patients. The general public requires spending additional expenses for medical services-related diseases. The patient pays for it relying upon how medical care is financed. Most times, extreme medical care related diseases lead to horribleness like all other contaminations. According to reports, paces of neonatal diseases and death rates among clinic conceived children in non-industrial nations have been multiple times higher than among medical clinic conceived infants in big league salary countries [31].

The expenses of medical services-related diseases were assessed at US dollar 7,00,000 every year in Trinidad and Tobago. Instalment for their treatment is made by dominant part of patients straightforwardly without help from anyone else. The costs paid by the patients and their families stay undetectable. Medical services-related diseases influence everybody, present and future patients. Treatment of medical care related contaminations is to be made frequently with anti-microbials, constraining the utilization of anti-toxins to increment and add to the endless loop of anti-toxin use and the development of antimicrobial resistance [31, 32].

A restrained and organized methodology is needed to plan and create clothing items. Prior to showing up at a plan arrangement, there ought to be a compelling and coordinated methodology through examination of the plan issue, depiction of plan prerequisites and basic investigation of those necessities. This could be cultivated through four significant pushes. Plan investigation and material examination are the initial two pushes, start the cycle in a way predictable with Rajendran and Anand description [33]. The last two pushes are stoop advancement and assessment.

Plan improvement and assessment, the last two pushes, move towards the plan arrangement. Figure 1 shows the stylish and capacity outrageous of items. Making a useful attire item tending to all pertinent prerequisites is a convoluted endeavour for which typifying the plan cycle is required. All clothing things should meet negligible useful necessities including body backing and stipend of some level of body development. Specialists and attendants wear careful covers in the working performance centre to address a double capacity of forestalling move of microorganisms and body liquids from the working staff to the patient and furthermore from patient to staff [34].

Careful covers should repulse sicknesses and diseases and furthermore give satisfactory opportunity to move, permit important versatility without scouring and abrading and should oppose tearing and linting. Without limiting development, they should fit intently. The covers should withstand steady pulls on the texture during routine developments as there is for the most part abundance texture. As medical clinics will just stock restricted amounts, these veils should be intended to fit a variety of body shapes with a restricted scope of sizes. The veils should help in keeping up the sterile zone needed for patient security and control the microscopic organisms delivered into the theatre. They ought not to have openings where the

Fig. 1 Aesthetic—functional extreme

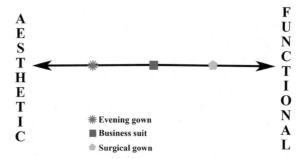

hindrance may be penetrated and ought to accommodate simple wearing and doffing without pollution. Single or multi-use veils should be adequately sturdy to last the planned valuable existence of the article of clothing. They ought to repulse liquids yet ventilate the specialist's extraordinary body heat [35].

The body's inside digestion for creation of warmth, general degree of action, outer temperature, protecting/penetrable properties of apparel to dampness, water or air and so on is the elements that add to warm solace. The centre temperature of human body stays 37 °C. Because of the digestion of food and muscle action, the body continually creates warmth, and this warmth is lost to climate. The warmth stream rate relies upon the texture properties and the temperature distinction; the more prominent the distinction the snappier warmth is changed. This property is known as temperature slope. Conduction, convection, and radiation are the three different ways by which warmth is moved from human body. At the point when the skin temperature is 33 °C, the body is in a condition of solace, and no fluid sweat is available. Warm solace is the capacity of the texture to keep a skin temperature and permit move of sweat created from the body. Warm solace properties of textures are controlled by the warm conductivity of the fabrics [36].

The solace wonder is affected by moisture recover likewise, as expansion in dampness recapture expands warm conductivity. Assurance of the general warm transmission coefficients because of the consolidated activity of conduction, convection, and radiation is covered by testing of warm conductivity. It quantifies the time pace of warmth move from a warm, dry, steady temperature and even level plate up through a layer of the test material to a generally quiet, cool air. Under pressure, the human body without much of stretch overheat and warmth feel is referred as complete warmth misfortune (THL). THL is an estimation of an article of clothing's capacity to permit body dampness and warmth to escape through the entirety of its layers. Simply 9–10 °F is the reach at which the centre internal heat level is 'typical' versus 'in danger of death'. The human mind can be influenced with additional progressions as low as 3–4 °F, and helpless dynamic can bring about extraordinary weakness, fatigue, loss of solidarity and sickness. Higher the all-out heat loss of the apparel framework, the worker in danger will have the more noteworthy physiological advantage. At the point when the labourers are more agreeable, they will actually want to work securely and adequately for a more extended time [37].

The human body is shielded by breathable textures from outer warmth, wind, water and numerous destructive specialists. Breathable textures license successful transmission of sodden fume from inside to outside climate. From the notable four climate recreation attire to particular clinical and military uses, the applications may go. Despite the fact that the water fume is permitted by breathable textures to diffuse through them, still they forestall the entrance of fluid water. The capacity of dress to permit the transmission of dampness fume by dissemination is the breathability of texture it encourages evaporative cooling. Meaning of breathability is frequently mistaken for wind entrance or dress' capacity to wick fluid water away from the skin. Both of these cycles are additionally alluded to as breathability, however, it is fundamentally relying upon totally extraordinary texture properties [38].

The nonwoven composite texture gives a boundary to blood, viral difficulties and breathability for comfort. Specifically, the texture is appropriate for use as a dispensable careful veil. The texture contains a first handle involving a miniature, permeable and formable gum covered into a non woven material substrate have extended to confer miniature porosity and in any event one extra employ situated neighbouring the main miniature permeable utilize. The nonwoven composite texture has obstruction properties passing the ASTM F1671 viral boundary test.39. It is fundamental that clinical professionals, medical attendants and others do not come into contact with the patient's blood because of revelation of AIDS. Specialists' veils should be impermeable to microorganisms and blood and furthermore should be agreeable. About 80% of clinical defensive dress in the USA is produced using nonwoven materials and utilized just a single time. For fluid repellency, the external surface of nonwoven is treated with fluorocarbon. The nonwoven is overlaid to a waterproof film in circumstance where an absolute boundary is required. At times, waterproof breathable movies are likewise utilized. In the UK and Europe, reusable defensive garments' are delivered from firmly woven cotton and polyester/cotton textures which are washed at high temperatures and sterilized [39].

A breathable, generously fluid impermeable cover, which is forever comparable to the forms of a wearer's body while being utilized in an individual consideration permeable article or a clinical article, is created. A breathable film and a nonwoven web are remembered for the cover. Endless supply of an extending power, both the movie and the web are extendible a cross way to a width of in any event 25% more prominent than the first width. Utilizing the overlay, a diaper or other article of clothing can be developed in modest style, bringing about material investment funds. At the point when the article of clothing is worn, to give a generously amazing fit on the wearer, the overlay extends just when required. Polyester/cotton mixes have been progressively utilized as covering substrates, due to resuscitated revenue in the 'customary' cotton, along with nylon textures which are less cruel in appearance and taking care of contrasted with ordinary fibre woven fabrics.

Cotton and polyester are conventional materials for careful covers which are agreeable to wear, yet the pore size is adequately huge to permit liquid or infections to go through; accordingly, there would not be obstruction impact until or except if some extraordinary measures are taken. Indeed, even rehashed impregnation with hydrophobic specialists on these materials, there is no assurance that it would be infective verification with wanted solace level. Firmly woven materials with a proper mix of polyester and long-staple cotton treated with fluid anti-agents fluorocarbon are highly preferable. After each wash, the anti-agents finish decreases and should routinely be reapplied to keep an obstruction impact.

Careful veils chose for use ought to be made out of an ignition safe material that sets up a successful boundary. Clean careful covers are made of either expendable single use nonwoven materials or reusable materials. As both of these sorts of materials, plan and execution attributes change, the varieties come from compromises in expenses, comfort, and the measure of obstruction assurance are given. The choice of cover material selection relies upon the proposed careful/intrusive strategy, and the level of blood and body liquid sprinkling is expected during the procedure [40].

For the determination of careful cover and veils, the accompanying practices are suggested by the Association of Operative Registered Nurses [41]:

a. Safe, distinguished requirements, and to advance patient and individual well-being are the variables deciding determination of material for development of careful veils and window hangings.
b. Selection of cover and wrap items for use in the training ought to be founded on rules explicit to the items' capacity and use. Careful covers and curtains are built of either single use or reusable materials.
c. Materials utilized for careful veils and curtains ought to be impervious to entrance by blood and other body liquids as required by their planned use.
d. Surgical veils and window hangings ought to keep up their uprightness and durability. Careful veils and curtains ought to have a worthy degree of standard, liberated from openings and deformities.
e. Abrasions, tears and penetrates safe properties are needed for careful veils and window hangings. The failure of careful veils to withstand tears, penetrates and scraped spots may consider entry of microorganisms, particulates and liquids among sterile and non-sterile territories and open patients to exogenous life forms.
f. Visual review of reusable materials ought to be made to decide their uprightness before use. Reusable materials ought to be fixed with heat-fixed patches of a similar quality material.
g. Surgical covers and curtains ought to be of low build-up. Boundary materials utilized for careful covers and window hangings ought to be as build-up free as could be expected.
h. Seams of boundary materials ought to be assessed to decide their ability to limit the infiltration or entry of expected pollutants. Numerous careful veil creases have almost no obstruction property.

i. Materials utilized for careful veils and curtains ought to be suitable to the techniques for sanitization. To accomplish a sterile item, the disinfecting specialist should contact all surfaces of the thing to be sanitized. Regular sanitization systems incorporate radiation, steam and ethylene oxide.

j. Tightly woven reusable material lose their defensive boundary quality after continued handling. The composed directions gave by the maker to dealing with, number of processing's, and the helpful existence of boundary materials ought to be followed. Careful covers and window hangings ought to be agreeable and contributable to keep up the wearer's ideal internal heat level. The careful items ought to have an alluring expense to profit proportion. Single utilize careful veils and curtains ought not to be re-disinfected except if producers give explicit composed guidelines to reprocessing.

Sicknesses to their hosts are brought about by irresistible specialists who are organic specialists. In medical care settings, numerous irresistible specialists are available. The most probable wellsprings of irresistible specialists and furthermore the most well-known vulnerable hosts are patients and medical services labours. Guests and labours in medical services may likewise be in danger of both disease and transmission. Contamination requires three primary components like wellspring of the irresistible specialist, method of transmission and a powerless host [42] (Fig. 2).

Irresistible specialists in medical services settings can be sent by contact, droplet and airborne. At the point when the exchange of microorganisms results from direct actual contact between a tainted or colonized individual and a helpless host, direct transmission happens. HCWs defiled hands contacting a weak site (like an injury)

Fig. 2 Chain of infection

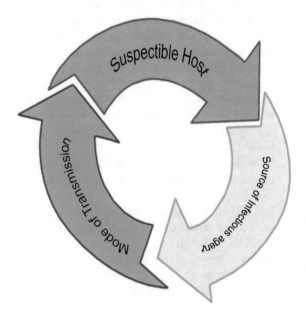

on a patient are a model. The detached exchange of an irresistible specialist to a defenceless host by means of a middle item is roundabout transmission. Instruments, bed rails over bed tables and other natural surfaces are the instances of transitional items. At the point when respiratory beads produced by means of hacking, sniffling or talking contact defenceless mucosal surfaces, like the eyes, nose or mouth, the drop transmission happens. Through contact with sullied structure with hands and afterwards mucosal surfaces, transmission may likewise happen in a roundabout way. Respiratory drops are normally scattered over brief distances as they are enormous and cannot stay suspended noticeable all around. Irresistible specialists that are spread by means of bead cores build-up from vanished drops containing infective microorganisms are eluded as airborne transmission. For significant stretches of time, these organic entities can get by outside the body and stay suspended noticeable all around. They contaminate others by means of the upper and lower respiratory tracts [42].

The utilization of careful veil appeared to shield the patients from diseases from the careful colleagues. The major prerequisite is to have a moderately inexactly woven, promptly porous and agreeable cotton material considering the rise of risks related with the transmission of blood-borne microbes, and the motivation unexpectedly has changed from ensuring the patient tainted by the specialist as well as to shield the specialist and paramedical staff from the contamination of the patients as well. Along these lines, the careful cover ought to shield the blood-borne irresistible microorganisms from entering through the texture, and the texture should act as a permanent proof [43].

Protective clothing for biological hazards is used in healthcare sector. Law enforcement and mortuary workers are in need of this protective clothing as they need to protect themselves from various hazards in the environment. Among the major causes of death, the increased morbidity among hospitalized patients is due to infections present in healthcare settings which are burden to both patient and for public health [44]. Hence, tri-laminate surgical mask design has been taken up for research to meet the global compliance and for high protection. Protective clothing plays an essential and meaty role in providing physiological comfort and also in protecting the human body. This research work aims to develop three layered fabric by combining polyester, polytetrafluoroethylene and viscose nonwoven. Analyse the suitability of these three layered fabric for antiviral surgical mask application to prevent from COVID-19.

2 Methods

Spun-bond polyester nonwoven fabrics with basis weight of 25 g/m^2 (GSM) were obtained from mogul nonwoven industry, Turkey. The diameter of the individual filament was approximately 23 μm. The basis weight of the sourced viscose nonwoven was 25 GSM. The thickness of the polyester nonwoven fabric was 0.26 mm. The pore size of polytetrafluoroethylene (PTFE) film was 0.1 μm. For the

Table 1 Fabric specifications

Specifications	Polyester	Viscose
Areal density (g/m^2)	25	25
Thickness (mm)	0.26	0.25
Individual filament fineness (μm)	0.8	6

development of three layered surgical mask, polyester was the outer layer, poly-tetrafluoroethylene film was used as middle layer and inner layer as a viscose nonwoven.

Fabrics were characterized with weight and thickness following ASTM D1777–64 standard method for measuring thickness of textile materials and ASTM D3776-85 standard method for mass per unit area of a fabric. The fabric characteristics of nonwoven fabrics are shown in Table 1.

2.1 Nonwoven Fabric Pore Size Characterisation

The image system used consists of a WIRA XEC-T20 trinocular compound microscope and a Moticam CCD camera mounted on the top of the microscope for capturing images and a true vision image board installed in PC for digitizing the images. A nonwoven fabric was placed in the microscope, and an objective of 400× magnification was selected to resolve images with a magnification about 473 pixels/mm on the computer monitor. The pore size of the nonwoven fabric was also analysed using capillary flow porometer according to standard ASTM F 316. In porometry principle, the liquid reduces the free energy of the system by spontaneously filling the pores of the solid, when the liquid–solid interfacial free energy is less than the solid gas interfacial free energy. These liquids, which are called as wetting liquids, cannot spontaneously come out of pores.

2.2 Development of Surgical Mask

Three layers of fabrics such as polyester nonwoven as an outer layer, PTFE film as a middle layer and viscose nonwoven fabric as an inner layer were bonded together using a fusing machine at a temperature of 210 °C with heated roller temperature of 240° and roller pressure of 120 N/cm^2. Figure 3 shows the fragmentary top view of laminated fabric in accordance with present research work.

Fig. 3 Three layered surgical
mask fragmentary view

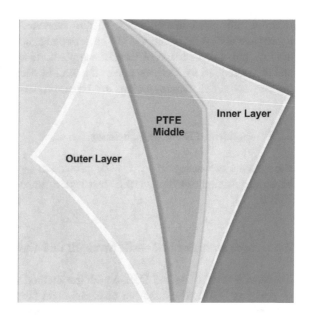

2.3   Testing Methods of Surgical Mask

2.3.1   Viral Penetration Test

Viral penetration testing of surgical mask using ASTM F 1671 was carried out to
investigate the antiviral property. ASTM F1671 is the test method used to measure
the resistance of materials used in protective clothing for penetration by
blood-borne pathogens, using bacteriophage under the condition of continuous
liquid contact. The test system has been designed for measuring penetration of
surrogate microbe for hepatitis B virus (HBV), hepatitis C virus (HCV) and HIV.
The protective clothing materials to be tested are intended to provide protection
against blood, body fluids and other potentially infectious materials. The surface
tension range for blood and body fluids is approximately 42–60 dynes/cm. In order
to simulate the wetting characteristics of blood and body fluids, the surface tension
of the ΦX174 bacteriophage suspension is adjusted to approximate lower end of
this surface tension range (42 dynes/cm).

The ΦX174 bacteriophage was selected as the most appropriate surrogate for the
blood-borne pathogens mentioned because it satisfies all of these criteria. The
ΦX174 bacteriophage is a non-enveloped 25–27 nm virus (similarly to HCV, the
smallest pathogen) with an icosahedral or nearly spherical morphology and
excellent environmental stability. It is non-infectious to humans, has a limit of
detection which approaches a single virus particle, grows rapidly and can be cultivated to reach high titers similar to HBV (the most concentrated pathogen mentioned). Test samples are prepared by randomly cutting the protective clothing

material into approximately 75×75 mm swatches. Test samples are exposed to approximately 60 ml of the $\Phi X174$ bacteriophage suspension. At the conclusion of the test, the observed side of the test sample is rinsed with a sterile assay medium and then analysed for the presence of the $\Phi X174$ bacteriophage. Surgical protective clothing 'pass/fail' determinations are based on detection of penetration.

2.3.2 Bacterial Filtration Efficiency

Face masks are designed to provide protection against biological aerosols. The bacterial filtration efficiency (BFE) test is performed on mask according to ASTM F2100.

2.3.3 Measurement of Tensile Strength and Elongation

The tensile strength ASTM D5034 grab test method was used for the determination of breaking load and elongation of tri-laminate fabrics since the procedures have been used extensively in the trade for acceptance testing. Tensile tests of the tri-laminate fabrics were performed using a Zwick Universal Testing Machine with a load cell of 10 kN capacity. Nonwoven fabric strip was cut into the dimensions 100×150 mm. The gauge length was set 80 mm, and each specimen was elongated at a crosshead speed of 50 mm/min.

2.3.4 Measurement of Tearing Strength

For the determination of tearing strength by the falling-pendulum type apparatus, ASTM D1424 test method is used. This test method, using a falling-pendulum Elmendorf type apparatus, covers the determination of the force required to propagate a single-rip tear starting from a cut in a fabric.

2.3.5 Measurement of Hydrostatic Resistance

To measure the resistance of a fabric to the penetration of water under hydrostatic pressure, the AATCC 127 method was used. Three specimens of fabric were diagonally taken across the width of the fabric to represent the material. The specimens were cut into 200×200 mm swatches to allow proper clamping. Hydrostatic pressure was applied on one surface of the test specimen, increasing at a constant rate, until three points of leakage appear on its other surface. At the top of inclined stand, one side of the specimen was clamped with the spring clamp, and to the free end of sample, another clamp of 0.4536 kg was fixed. In the funnel of the tester, 500 ml of distilled water was poured, and onto the specimen, it was allowed to spray from a height of 60 cm. Water pressure is applied on a test specimen mounted under orifice of a conical well, increasing at a constant rate until three leakage points appear on its surface.

2.3.6 Measurement of Moisture Vapour Transfer Rate (MVTR)

The standard method used to test the MVTR is ASTM E96. The common methods of tests are the upright cup, inverted cup and desiccant methods. The range of humidity can be 50%, and temperature can be 70 °F or 90 °F. Depending upon the type of breathable system, the test figures vary widely. Microporous coatings have excellent upright and desiccant cup values. With the inverted cup, much higher transmission rates can be produced by the hydrophilic membrane. Manufacturers tend to adopt testing methods that suit to their system [45].

To test the characteristics of the breathability of fabrics, moisture vapour transfer rate measurement is made. To evaluate the MVTR of three layered nonwoven mask fabric, evaporative dish method following the British Standard BS 7209:1990 was used. The testing was carried out at standard laboratory conditions of $65 \pm 2\%$ relative humidity and 20 ± 2 °C temperature [46].

3 Results

The smallest pore diameter, mean flow pore diameter and bubble point diameter of surgical mask are reported in Table 2. Mean flow pore diameter of three layered surgical mask is higher than polypropylene nonwoven mask due to polyester nonwoven fabric comprises coarser individual filament fineness and more void of viscose nonwoven than polypropylene fabric. The increase in mean flow pore diameter has increased the moisture vapour permeability. The pore size of polyester nonwoven is 0.33 μm. Mean flow pore diameter of the dual laminated polyester nonwoven with PTFE surgical mask is reduced due to the presence of stretched PTFE polymer matrix.

Table 2 Pore size of single, dual and three layered surgical mask

Surgical mask	Smallest pore diameter (μm)	Mean flow pore diameter (μm)	Bubble point diameter (μm)
Single layer polyester surgical mask	0.31	0.33	0.52
Dual layer of polyester and PTFE film surgical mask	0.16	0.30	0.48
Three layered surgical mask (polyester/PTFE/viscose)	0.19	0.39	0.44

Table 3 Results of (polyester/PTFE/viscose) surgical mask

Sample	Viral penetration	Tensile strength (N)		Tearing strength (N)		Bacterial filtration efficiency	Hydro static resistance (cm)
		MD	CD	MD	CD	(%)	
Single layer polyester surgical mask	Fail	68	41	5	1.5	73.5	22.3
Dual layer of polyester and PTFE film surgical mask	Pass	105	62	7	2	98.7	31.2
Three layered surgical mask (polyester/PTFE/viscose)	Pass	190	108	8	3	99.4	37.6

Table 3 shows the results of mechanical properties of single, dual and three layered surgical mask. The developed three layered fabric with polyester as an outer layer, microporous PTFE film as a middle layer and viscose nonwoven fabric as an inner layer, named as PET 1 mask, has passed the viral penetration test, which confirms that the mask has protection against the hepatitis B and hepatitis C, human immunodeficiency viruses, coronavirus and surrogate ΦX174 bacteriophage. In single layer mask, one correlate with level 1 mask protection, whereas dual, three layered mask having high level of protection with bacterial filtration efficiency of greater than 95% (Fig. 4).

It is observed from Fig. 2 that while laminating with PTFE, polyester and viscose nonwoven fabrics, the water vapour permeability reduces as the layer increases. The moisture vapour permeability of single layer polyester mask is 985 g/m^2/day which is better than single layer polypropylene because the pore size of polyester is 0.33 μm which is higher than polypropylene pore size of 0.29 μm.

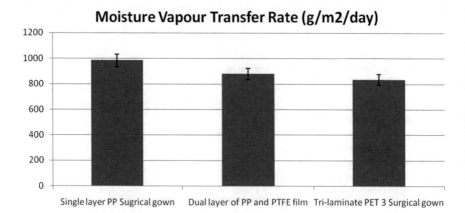

Fig. 4 Water vapour permeability single, dual and three layered mask

4 Discussions

Single layer polyester nonwoven surgical masks failed in viral penetration analysis because the pore size of polyester was 0.33 μm, and the thickness of polyester nonwoven was 0.26 mm. Single layer polyester nonwoven surgical mask failed viral penetration test as the average pore diameter is 0.33 μm which is more than size of ΦX174 bacteriophage virus as 0.2 μm and in the absence of PTFE film. Dual layer of polyester and PTFE film surgical mask passed the viral penetration analysis, as the PTFE has 0.1 μm porosity through which viruses cannot penetrate. Hence, it is fused with an outer polyester layer to give liquid barrier protection. Single layer polyester nonwoven surgical mask failed viral penetration test as the average pore diameter is 0.33 μm which is more than size of virus and in the absence of PTFE film.

Dual and three layered PET 1 surgical mask passed viral penetration analysis as it has smallest pore diameter is 0.16 and 0.19 μm, respectively. Polyester nonwoven fabric is used as an outer layer because it is hydrophobic fibre and does not attract water. It is also non-polar and hence repels blood fluids. Higher surface area of the microfibers in the spun-bond polyester nonwoven fabric, which is the outer layer, acts as a liquid barrier.

The tensile strength of the PET 1 three layered fabric in machine direction (MD) is 190 N which is greater than the cross direction (CD) of the sample which is 108 N. Paired t-test analysis of the tensile strength in machine and cross directions shows that there is significant difference at 95% confidence interval. The tearing strength of the three layered mask fabric in direction I and II was 8 and 3 N, respectively. The tearing strength of the fabric in the direction I is more, due to the laying of polyester fibre in the machine direction during extrusion. In hydro static resistance, it is observed that the weight of water penetrating through the fabric during impact penetration test decreases as the weight of fabric increases. The results are statistically significant for single layer and PET 1 three layered mask at 95% confidence level.

The PET 1 mask moisture vapour permeability test result was 834 g/m^2/day which also confirm the mask requirements. Spun-laced viscose nonwoven absorption influences the better moisture transfer. PTFE film allows water vapour to pass through. The water vapour from perspiration be transmitted from inside to outside so that under do not become wet, and a natural evaporative cooling effect can be achieved. Though the addition of viscose nonwoven as the inner layer increases moisture vapour permeability and drying rate of the surgical mask. The developed three layered mask can be used for protecting healthcare people from the risk of COVID-19 and high risk of exposure to fluid amount, splash and pressure on mask in medical procedures.

The surgical mask has better moisture vapour transfer rate as compared to other commercial polypropylene mask. So that wearer can use it with comfort as the inner layer gives soft touch to wearer's skin increasing the property of comfort to the user. In the current challenging situation, the mask having coronavirus barrier property is also established in this surgical mask thereby reducing cross-contamination between medical staff and patients.

5 Conclusions

Polyester nonwoven fabrics have pore size of 0.33 μm, and PTFE membrane pore has 0.1 μm thereby providing better protection against COVID-19. Developed single layer polyester mask confirmed level 1 protection. PET 1 surgical mask confirmed level 4 protection as per AAMI barrier classification. Three layered polyester and dual laminate polyester mask passed viral penetration, whereas single layer polyester mask failed in viral penetration analysis. Because of the absence of PTFE membrane, the challenge liquid suspension passes over the fabric in the single layer polyester surgical mask. PTFE membrane passes moisture vapour but prevents virus penetration. PET 1 three layered polyester mask possesses higher tensile and tearing strength, due to strength of polyester, extensibility of PTFE and increase in the number of layers. As per hydrostatic resistance findings, the PET 3 mask offers level 2 protection. The hydrostatic resistance increases with increase in weight of the fabric in all masks.

The PET 3 surgical mask has better moisture vapour transfer rate, so that wearer can use it with comfort as the inner layer gives soft touch to wearer's skin increasing the property of comfort to the user. Liquid barrier property is also established in this surgical mask, thereby reducing cross-contamination between medical staff and patient. The developed PET 3 three layered mask can be used for high risk of exposure to COVID-19, splash and pressure on mask such as procedures like open cardiovascular, Trauma procedures, Cesarean sections and any procedure in which the surgeon is prone to cavity.

References

1. Rowan NJ, Laffey JG (2020) Challenges and solutions for addressing critical shortage of supply chain for personal and protective equipment (PPE) arising from coronavirus disease (COVID19) pandemic—case study from the Republic of Ireland. Sci Total Environ. https://doi.org/10.1016/j.scitotenv.2020.138532
2. Swennen GRJ et al (2020) Custom-made 3D-printed face masks in case of pandemic crisis situations with a lack of commercially available FFP2/3 masks. Int J Oral Maxillofac Surg 49 (5):673–677
3. Akladios C et al (2020) Recommendations for the surgical management of gynecological cancers during the COVID-19 pandemic—FRANCOGYN group for the CNGOF. J Gynecol Obstet Hum Reprod. https://doi.org/10.1016/j.jogoh.2020.101729
4. Wu et al (2020) Facemask shortage and the novel coronavirus disease (COVID-19) outbreak: reflections on public health measures. EClin Med. https://doi.org/10.1016/j.eclinm.2020.100329

5. Swennen GRJ et al (2020) Custom-made 3D-printed face masks in case of pandemic crisis situations with a lack of commercially available FFP2/3 masks. Int J Oral Maxillofac Surg. https://doi.org/10.1016/j.ijom.2020.03.015

6. Dane JH, Schwartz P (2005) Effect of liquid characteristics on the wetting, capillary migration and retention properties of fibrous polymer networks—part 2. J Appl Polym Sci 98:384–390

7. Mews P (2009) Establish and maintain the sterile field competency for safe patient care during operative and invasive procedures. Denver 1:240–249

8. Loveday HP, Wilson JA, Hoffman PN, Pratt RJ (2007) Public perception and the social and microbiological significance of uniforms in the prevention and control of healthcare-associated infections: an evidence review. Br J Infect Control 8:10–21

9. Lin H, He N, Su M, Feng J, Chen L, Gao M (2011) Herpes simplex virus infections among rural residents in eastern China'. BMC Infect Dis 11:1–6

10. Association of Operating Room Nurses (AORN) (2010) Recommended practices: protective barrier materials for surgical masks and drapes. AORN J 55:832–835

11. Neely AN, Maley MP (2000) Survival of Enterococci and Staphylococci on hospital fabrics and plastics. J Clin Microbiol 38:724–726

12. Burgatti JC, Lacerda RA (2009) Systematic review of surgical masks in the control of contamination/surgical site infection. Rev Enferm 43:229–236

13. Ducel G, Fabry J, Nicolle L (2002) Prevention of hospital acquired infections. World Health Organisation, France

14. Huang W, Leonas K (2000) Evaluating a one-bath process for imparting antibacterial and repellency to nonwoven surgical mask fabrics. Text Res J 70:774–782

15. Belkin NL (2002) Barrier surgical mask and drapes: just how necessary are they. Text Rental 1:66–73

16. Association for the Advancement of Medical Instrumentation (AAMI) (2005) Selection and use of protective apparel and surgical drapes in healthcare facilities. AAMI Tech Inf Rep 1:12–24

17. Okeke IN, Laxminarayan R, Bhutta ZA, Duse AG, Jenkins P, Obrien TF (2005) Antimicrobial resistance in developing countries. Part I: recent trends and current status. Lancet Infect Dis 5:481–493

18. Fung W (2002) Coated and laminated textiles. Wood Head Publishing, England

19. Johnson L, Schultze D (2000) Breathable TPE films for medical applications. AORN J 10:50–53

20. Wang C, Horby PW, Hayden FG, Gao GF (2020) A novel coronavirus outbreak of global health concern. Lancet 395(10223). https://doi.org/10.1016/S0140-6736(20)30185-9

21. World Health Organization (WHO) (2013) Prevention of hospital-acquired infections—practical guide. Commun Dis Surveill Response 1:5–40

22. Mangram AJ, Horan TC, Pearson ML, Silver LC, Jarvis WR (1999) Guideline for prevention of surgical site infection. Infect Control Hosp Epidemiol 20(4):250–278

23. Nurmi S, Lintukorpi A, Saamanen A, Luoma T, Soinnen M, Suikkanen V (2003) Human, protective cloths and surgical drapes as a source of particles in an operating theatre. Autex Res J 3(1):394–399

24. Scott R (2005) Textiles for protection. Woodhead Publication, England

25. Unsal E, Dane JH, Schwartz P (2005) Effect of liquid characteristics on the wetting, capillary, migration and retention properties of fibrous polymer networks. J Appl Polym Sci 97(1):282–292

26. Wong ES, Rupp ME, Mermel (2005) Public disclosure of healthcare-associated infections: the role of the Society for Healthcare Epidemiology of America. Infect Control Hosp Epidemiol 26(1):210–212

27. Occupational Safety and Health Administration (1989) Occupational exposure to blood borne pathogens: proposed rule and notice of hearing. US Department of Labor Federal Register, pp 3042–3139

28. European Committee for Standardisation (2012) European standard for surgical drapes and masks. European Standard, Luxemboug

29. European Centre for Disease Prevention and Control (2011) European antimicrobial resistance surveillance network, Sweden
30. Zaidi AK, Huskins WC, Thaver D, Bhutta ZA, Abbas Z, Goldmann DA (2005) Hospital acquired neonatal infections in developing countries'. Lancet 365(3):1175–1188
31. Raza MW, Kazi BM, Mustafa M, Gould FK (2004) Developing countries have their own characteristic problems with infection control. J Hosp Infect Control 57(4):294–299
32. Campbell SL (1996) Creative approaches to PPE development. Occup Health Saf J 12(5):82–91
33. Rajendran S, Anand SC (2002) Development of medical textiles. The Textile Institute, Manchester
34. Slater K (1998) Textile use in surgical mask design. Can Text J 115(4):16–18
35. Belkin NL (1993) The challenge of defining the effectiveness of protective aseptic barrier. Tech Text Int J 1:22–24
36. Dent RW (2001) Transient comfort phenomena due to sweating. Text Res J 71(9):796–806
37. Mitu S, Potoran I (1971) Fundamentals of textile clothing technology. Bull Inst 17(1):61–70
38. Langley JD, Hinkle BS, Carroll TR, Vencill CT (2004) Breathable blood and viral barrier fabric extracts from european patent applications part-1B: primary industry fixed constructions mining. European patent 1448831
39. Mukhopadhyay A, Vinay Kumar M (2008) A review on designing the waterproof breathable fabrics part II: construction and suitability of breathable fabrics for different uses. J Ind Text 38(1):17–41
40. Snycerski M, Frontczak I (2004) A functional woven fabric with controlled friction coefficients preventing bedsores. AUTEX Res J 4(30):137–142
41. Association of Operating Room Nurses (AORN) (2010) Recommended practices: protective barrier materials for surgical masks and drapes. AORN J 55(3):832–835
42. Lin H, He N, Su M, Feng J, Chen L, Gao M (2011) Herpes simplex virus infections among rural residents in eastern China. BMC Infect Dis 11(69):1–6
43. Jessie H, Darlene M, Doris H (2007) Antibacterial and laundering properties of AMS and PHMB as finishing agents on fabric for healthcare workers. Uniforms' Clothing Text Res J 25 (3):258–272
44. Datta SD, Armstrong GL, Roome AJ, Alter MJ (2003) Blood exposures and hepatitis C virus infections among emergency responders. Arch Intern Med 163(21):2605–2610
45. Junyan H, Li Y, Kwokwing Y, Wong Anthony SW, Weilin X (2005) Moisture management tester: a method to characterize fabric liquid moisture management properties. Text Res J 75 (1):57–62
46. McCullough EA, Kwon M, Shim H (2003) A comparison of standard methods for measuring water vapour permeability of fabrics. Meas Sci Technol 14(8):1402–1408

Printed in the United States
by Baker & Taylor Publisher Services